D1361525

Main lectures presented at the

Xth and XIth MICROSYMPOSIA ON MACROMOLECULES

Conférences principales présentées aux

Xᵉ et XIᵉ MICROSYMPOSIA SUR LES MACROMOLECULES

UNION INTERNATIONALE DE
CHIMIE PURE ET APPLIQUEE

DIVISION DE CHIMIE MACROMOLECULAIRE

en relation avec

L'ACADEMIE TCHECOSLOVAQUE DES SCIENCES

et

LA SOCIETE CHIMIQUE TCHECOSLOVAQUE

MICROSYMPOSIA MACROMOLECULAIRES —X et XI

Conférences principales présentées aux

Xᵉ et XIᵉ MICROSYMPOSIA SUR LES MACROMOLECULES

à Prague, Tchécoslovaquie
28 août–7 septembre 1972

Rédacteur du Symposium
B. SEDLÁČEK

LONDRES

BUTTERWORTHS

INTERNATIONAL UNION OF
PURE AND APPLIED CHEMISTRY

MACROMOLECULAR DIVISION

in conjunction with

CZECHOSLOVAK ACADEMY OF SCIENCES

and

CZECHOSLOVAK CHEMICAL SOCIETY

MACROMOLECULAR
MICROSYMPOSIA—X and XI

Main lectures presented at the

Xth and XIth MICROSYMPOSIA ON MACROMOLECULES

at Prague, Czechoslovakia
28 August–7 September 1972

Symposium Editor
B. SEDLÁČEK

Distributed in the United States by
CRANE, RUSSAK & COMPANY, INC.
347 Madison Avenue
New York, New York 10017

ENGLAND: BUTTERWORTH & CO. (PUBLISHERS) LTD.
LONDON: 88 Kingsway. WC2B 6AB

AUSTRALIA: BUTTERWORTHS PTY. LTD.
SYDNEY: 586 Pacific Highway, Chatswood, NSW 2067
MELBOURNE: 343 Little Collins Street, 3000
BRISBANE: 240 Queen Street, 4000

CANADA: BUTTERWORTH & CO. (CANADA) LTD.
TORONTO: 14 Curity Avenue. 374

NEW ZEALAND: BUTTERWORTHS OF NEW ZEALAND LTD.
WELLINGTON: 26–28 Waring Taylor Street. 1

SOUTH AFRICA: BUTTERWORTH & CO. (SOUTH AFRICA) (PTY) LTD.
DURBAN: 152–154 Gale Street

The contents of this book appear in

Pure and Applied Chemistry, Vol. 36, Nos. 1–2 (1973)

Suggested U.D.C. *number:* 541·64 (063)

©

International Union of Pure and Applied Chemistry
1973

ISBN 0 408 70540 X

Printed in Great Britain by Page Bros (Norwich) Ltd.. Norwich

MAIN LECTURES

SCIENTIFIC AND
ORGANIZING COMMITTEE

Chairman: B. SEDLÁČEK
Members: D. DOSKOČILOVÁ
            S. NEŠPŮREK
            V. PETRUS
            H. PIVCOVÁ
            J. POSPÍŠIL
            B. SCHNEIDER
            L. TAIMR
            L. ZIKMUND

ON THE DOMINANCE OF SHORT-RANGE INTERACTIONS IN POLYPEPTIDES AND PROTEINS

HAROLD A. SCHERAGA

Department of Chemistry, Cornell University, Ithaca, New York 14850, USA

ABSTRACT

The development of the concept of the dominance of short-range interactions in polypeptides and proteins is traced. It is shown that certain amino acid residues have a preference to adopt the right-handed α-helical conformation while others exhibit a preference to participate in β-turns. These ideas are being applied in conformational energy calculations to try to determine the native conformations of proteins.

I. INTRODUCTION

In computing protein conformation from amino acid sequence, using empirical energy functions, the problem of the existence of many local minima in the multi-dimensional energy surface is encountered[1]. Since (energy) minimization algorithms lead to the nearest local minimum, depending on the starting point, it is desirable to find alternative methods that would lead to initial starting points which might have a reasonable chance of being in the desired potential well; then, energy minimization would lead to the desired local minimum. For reasons outlined in section III, one of the approaches for finding alternative methods led to a consideration of the possible dominance of short-range interactions[2, 3]. This investigation led to the concept[2, 3] that the conformation of an amino acid residue in a polypeptide or protein is determined in very large measure (though not exclusively) by the short-range interactions between a sidechain and the atoms of the backbone of the *same* amino acid residue, and is (again in first approximation) essentially independent of interactions with neighbouring sidechains or backbone portions of the chain. This view has recently received further support from a statistical analysis of the conformations of amino acid residues in globular proteins by Finkelstein and Ptitsyn[4]. In this paper, we shall trace the development and application of this concept.

II. DEFINITIONS

As used here, the term 'short-range' refers to an interaction between the sidechain of an amino acid residue with its own backbone. The interaction between the atoms of a given residue with those of any other residue, nearby

1

in the chain or more remote along the chain (even though, possibly, nearby in space) is termed 'long-range'.

III. THE θ-POINT

The treatment of an ideal homopolymer chain by random-flight statistics leads to the conclusion that some average linear dimension of the chain, e.g. the root-mean-square end-to-end distance, $\langle \overline{r^2} \rangle^{\frac{1}{2}}$, varies with the square root of the molecular weight[5]. While long-range and excluded volume effects (not included in the random-flight calculation) tend to increase $\langle \overline{r^2} \rangle^{\frac{1}{2}}$ beyond its *ideal* value, the choice of an appropriate (poor) solvent (in which polymer–polymer contacts are favoured over polymer–solvent contacts) can reduce $\langle \overline{r^2} \rangle^{\frac{1}{2}}$ to its *ideal* value[5]. Under these conditions (i.e. at the θ-point), the polymer–polymer and polymer–solvent interactions compensate the long-range and excluded volume effects, and the ideal value of $\langle \overline{r^2} \rangle^{\frac{1}{2}}$ which results is determined by short-range interactions[5]. Although a protein in aqueous salt solution may not be at the θ-point, the possibility existed that its conformation, while not determined exclusively by short-range interactions, might nevertheless be dominated by them. As will be shown here, the dominance of short-range interactions has been demonstrated for the formation of α-helical and non-helical portions of proteins[2–4] and for the formation of β-turns[6]; it remains to be seen whether β-structures also are determined in large measure by short-range interactions†.

IV. CONFORMATIONAL PREFERENCES WITHIN A SINGLE PEPTIDE UNIT

To examine the validity of the hypothesis that short-range interactions are dominant, a study was made[2] of the role of these interactions in helix formation for proteins of known structure. In particular, calculations were carried out to obtain the energy of interaction of individual sidechains in lysozyme with sidechains that are nearest neighbours along the backbone, as well as with the backbone groups themselves. It was found that, for various initial backbone conformations (viz. the right- and left-handed α-helices, α_R and α_L, respectively, and the antiparallel pleated sheet structure, β), the conformation of lowest energy after minimization was the same in most cases for a given amino acid residue and was independent of the nature of the next amino acid in the chain. Furthermore, the backbone structures corresponding to the lowest energy (i.e. α_R, β or α_L) showed a high degree of correlation with the so-called helix-making or helix-breaking character of a residue, as determined by earlier *empirical* studies on the identification of α-helical regions in proteins[7–10]. In other words, it appears that the short-range interactions within a given peptide unit may be the physical origin of the so-called helical potential of a residue. In addition, since the sidechain–

---

† Even though a parallel or anti-parallel β-structure involves hydrogen bonds (which are long-range interactions), nevertheless the possibility exists that certain residues have a preference for β-conformations because of short-range interactions, and the hydrogen bonds might then contribute added (long-range) stability.

sidechain interaction does not usually play a major role in determining conformation, the cooperativity among residues, which is necessary for the formation of a helical segment, may simply be the additive effect of placing some sequence of helix-making residues in a particular region. This suggested a model for helix formation in which each type of peptide unit in proteins of known amino acid sequence was assigned a designation h or c (helix-making or helix breaking, respectively), based on a study of the energy surface of the peptide unit. Then, from an examination of the h or c assignments for lysozyme, myoglobin, α-chymotrypsin and ribonuclease, empirical rules were formulated to distinguish between helical and non-helical regions. These rules are: (a) an α-helical segment will be nucleated when at least four h residues in a row appear in the amino acid sequence and (b) this helical segment will continue growing toward the C-terminus of the protein until two c residues in a row occur, a condition that terminates the helical segment. With these rules, it was possible to predict the helical or non-helical state of 78 per cent of the residues of the four proteins mentioned above[3].

With the later availability of the x-ray structures of seven proteins, the validity of these rules was examined further[11]. It was observed that, if a dipeptide ever occurred at the C-terminus of a helical region, it had a low probability of occurring elsewhere in a helical region and as high as a 90 per cent probability of occurring elsewhere in non-helical regions. It was also found that those residues designated as c tended to predominate at the C-termini of helical segments. These results constitute an experimental demonstration of the validity of rule b above. Finkelstein and Ptitsyn[4] also made a statistical analysis of the conformations of amino acid residues in proteins of known structure, and came to similar conclusions, viz. that short-range interactions are dominant, in that single residues can be classified as helix-making or helix-breaking and that sidechain–sidechain interactions play a minor role in determining the conformational preference of a given amino acid residue.

At this point, it is of interest to consider the factors which determine the conformational preference of a given amino acid residue. The conformational entropy of a residue in the random coil state must be overcome by favourable energetic factors in order for the residue to be helix-making; otherwise, it will be helix-breaking. Glycyl residues, with no sidechains, have no favourable energetic factors to enhance helix formation; thus, the entropy of the coil makes glycyl residues helix-breaking[12]. When a β-CH_2 group is added, the resulting non-bonded interactions tend to favour the α_R conformation[2, 12]. However, as in Asn which is helix-breaking, electrostatic interaction between a polar sidechain group and the polar backbone amide group de-stabilizes the α_R conformation relative to other conformations. In Gln and Glu, the electrostatic effect is weaker because of the greater distance between the backbone amide group and the polar sidechain group (resulting from entropically-favoured extended sidechain conformations); hence, the preferred conformation for Gln and Glu is α_R. Recently, an extensive series of conformational energy calculations (including the computation of statistical weights) was carried out for the N-acetyl-N'-methyl amides of all twenty naturally-occurring amino acids[13]. From these calculations, it is possible to assess how the various energetic factors contribute to the conformational preferences of each residue. For example, the sidechains of both Ser and Asp can form hydrogen bonds with the nearby backbone amide groups when these residues are in non-helical conformations; thus, Ser and Asp are helix-breaking.

V. QUANTITATIVE SPECIFICATION OF HELIX-MAKING AND HELIX-BREAKING CHARACTER

Having demonstrated that the conformation of an amino acid residue in a protein is determined largely by short-range interactions, and, thus, in first approximation is essentially independent of the chemical nature of its neighbours, it becomes desirable to have a quantitative scale to specify the helix-making and helix-breaking character of the twenty naturally-occurring amino acids—instead of the earlier[2, 3] assignment of all amino acids to two categories, h or c. A model which suggests itself is the helix–coil transition in homopolymers; i.e. the Zimm–Bragg parameters σ and s (cf. ref. 14), which characterize the transition curve, would appear to provide a quantitative basis for specifying the helix-making and breaking tendency of any amino acid in its corresponding homopolymer (and, therefore, in a protein, since short-range interactions dominate in both cases). Because of certain experimental problems discussed elsewhere[1, 15, 16], homopolymers cannot be used for this purpose, and resort is had instead to the use of random copolymers of two components—a helical host, for which σ and s are known, and the guest residues; from the effect of increasing amounts of the guest residues on the helix–coil transition curve of the homopolymer of the host residues, it is possible to determine σ and s for the guest residues[1, 15, 16]. Thus far, these experiments have been carried out for the following guest residues: Gly[17], Ala[18], Ser[19] and Leu[20], and the results are shown in *Table 1*. It can be seen that Gly and Ser are helix-breakers, Gly more so than Ser (because $s < 1$), and Ala and Leu are helix-makers, Leu more so than Ala. Since the experiments were carried out in aqueous solution[16–20], the resulting experimental values of σ and s contain all energetic and entropic contributions (including solvation) which determine the conformational preference.

Table 1. Experimental values of σ and s determined by the host–guest technique

Temp., °C	s			
	Glycine[a] (ref. 17)	L-Alanine[b] (ref. 18)	L-Serine[c] (ref. 19)	L-Leucine[d] (ref. 20)
0	0.51	1.08	0.73	1.10
10	0.55	1.08	0.77	1.12
20	0.59	1.07	0.78	1.14
30	0.62	1.06	0.79	1.14
40	0.63	1.04	0.79	1.13
50	0.63	1.02	0.78	1.11
60	0.63	1.01	0.74	1.09
70	0.61	1.00	0.72	1.06

[a] $\sigma = 1 \times 10^{-5}$; [b] $\sigma = 8 \times 10^{-4}$; [c] $\sigma = 7.5 \times 10^{-5}$; [d] $\sigma = 33 \times 10^{-4}$.

VI. HELIX PROBABILITY PROFILES

The experimental values of σ and s can be used to obtain information about the conformation of any specific sequence of amino acids, e.g. that of a protein. However, since the values of σ and s were obtained from the Zimm–Bragg theory, which is based on the one-dimensional Ising model, we cannot treat the *native* protein molecule since its conformation is, in some measure, influenced by long-range interactions which are not taken into account in the Zimm–Bragg theory. Since the *denatured* protein is devoid of tertiary structure and hence, presumably, of long-range interactions other than excluded volume effects, the polypeptide conforms to the one-dimensional Ising model. Thus, above the denaturation temperature, we may apply the Zimm–Bragg formulation to this copolymer of ~ 20 amino acids to determine the probability that any given residue of the chain will be in the α_R or in the random coil conformation, respectively[21]. It will then be shown that there is a correlation between the calculated α_R probability profile of the *denatured* protein and the experimentally observed α_R regions in the corresponding *native* structures; i.e. in many cases, those regions in the denatured protein which exhibit a propensity for being in the α_R conformation correspond to the α_R regions observed in the native protein.

The partition function Z, and the probability, $P_H(i)$, that the ith amino acid (of type A) in a chain of N residues is in the α_R conformation are given by

$$Z = (0, 1)\left[\prod_{j=1}^{N} W_A(j)\right]\binom{1}{1} \tag{1}$$

and

$$P_H(i) = (0, 1)\left[\prod_{j=1}^{i-1} W_A(j)\right]\frac{\partial W_A(i)}{\partial \ln s_A(i)}\left[\prod_{j=i+1}^{N} W_A(j)\right]\binom{1}{1} \bigg/ Z \tag{2}$$

where $W_A(j)$ is the matrix of statistical weights for the jth residue which is of amino acid type A, viz.

$$W_A(j) = \begin{pmatrix} s_A(j) & 1 \\ \sigma_A(j)s_A(j) & 1 \end{pmatrix} \tag{3}$$

$s_A(j)$ is the statistical weight assigned to this residue when it is in an α_R conformation and preceded by a residue in the α_R conformation, and $\sigma_A(j)s_A(j)$ is the statistical weight assigned to this residue when it is in an α_R conformation and preceded by a residue in the random coil conformation. The use of equation 2 to compute $P_H(i)$ automatically includes the cooperativity which is characteristic of the nearest-neighbour one-dimensional Ising model.

Pending the acquisition of data, such as those of *Table 1*, for the remainder of the twenty naturally-occurring amino acids, the set of amino acids has been grouped into three categories (all with σ taken as 5×10^{-4}), viz. helix-breakers (with $s = 0.385$), helix-formers (with $s = 1.05$), and helix-indifferent (with $s = 1.00$). Taking into account the limited data of *Table 1*, the earlier h and c assignments of Kotelchuck and Scheraga[2, 3], and the results of an information-theory analysis by Pain and Robson[22], the amino acids are assigned as in *Table 2*. It should be emphasized that these tentative values of

5

Table 2. Assignment of amino acid residues to three categories according to helix-forming power

Helix-breaker		Helix-indifferent			Helix-former		
Gly	Pro	Lys	Asp	Arg	Val	His	Met
Ser	Asn	Tyr	Thr	Cys	Gln	Ala	Leu
				Phe	Ile	Trp	Glu

σ and s are used here only pending completion of experiments which will complete *Table 1* for the remaining amino acids.

Helix probability profiles for eleven proteins have been calculated from equation 2, using the values of σ and s discussed above and the assignments of *Table 2*[23]‡. From these curves it appears that there is a close correlation between the propensity of a particular amino acid residue to be in the α_R conformation in the denatured protein and its occurrence in a helical region in the globular structure of the corresponding native protein. On this basis, it was suggested[21] that, during renaturation, the protein chain acquires *specific* long-range interactions which stabilize the helical regions which tend to form in certain portions of the chain; i.e. folding of the polypeptide chain into the native conformation of a protein is thought to occur by incipient formation of α-helical or other ordered structural regions (among those residues with a propensity to be helical) stabilized by specific long-range interactions, with the remainder of the protein molecule then folding around these stabilized helical regions.

Consistent with this view, it is found[23] that, despite amino acid substitutions in a series of 27 species of cytochrome c proteins, there is a striking similarity in their helix probability profiles, and a good correlation with the location of the helical regions in the (x-ray determined) structure of the horse and bonito proteins. It appears that amino acid substitutions may be tolerated in evolution, provided that the helix-forming or helix-breaking tendency (i.e. values of σ and s) of each amino acid residue is preserved, thereby enabling the altered protein to maintain the same three-dimensional conformation and, hence, the same biological function.

Application of this approach to lysozyme and α-lactalbumin[25], two different proteins with striking homologies in their amino acid sequence, led to very similar helix probability profiles. This result supports earlier suggestions[26–28] that the two proteins might have similar three-dimensional structures, and, if so, again demonstrates the conservative nature of amino acid replacements (as far as helix-forming power in homologous proteins is concerned) which was found for the cytochrome c proteins.

VII. β-TURNS

While the protein chain must fold in order to enable remote helical or other ordered structural regions to approach each other to be stabilized by long-range interactions, it is felt that these long-range interactions are not

‡ More detailed information about the conformational state of each residue is provided by a recently-developed eight-state model for the helix–coil transition in homopolymers and specific-sequence copolymers[12, 24].

brought into play by a *chance* encounter of the ordered regions[6]. Instead, there is a tendency for bends or β-turns to occur among certain amino acid residues, thereby 'directing' the encounter of the ordered regions. From a statistical analysis of the amino acid compositions of the bends in three proteins, it has been possible to formulate rules for the existence of β-turns in general. Application of these rules to several other proteins led to a high degree of correlation between the predicted regions where the β-turns should appear, and their existence in the (x-ray determined) structure[6]. A discussion of β-turns in proteins has also been presented by Kuntz[29]. It is of interest that residues like Gly, Ser and Asp, which have a low tendency toward helix formation, have a high propensity to form β-turns.

We have just completed a study[30] of the bends found in the native structures of eight proteins. The 135 bends which were located could be grouped among ten types, and over 40 per cent of the bends did not possess a hydrogen bond between the $C{=}O$ of residue i and the NH of residue $i + 3$. In addition, conformational energy calculations were carried out on three pentapeptides with amino acid sequences found as bends in the native structure of α-chymotrypsin. The results indicate that the bends occur not only in the whole molecule, but also in the pentapeptide; i.e. the observed bends were the conformations of lowest energy even in the pentapeptides. The stability of the bends, compared to those of other structures, arises principally from side-chain–backbone interactions (e.g. a hydrogen bond between the sidechain COO^- of Asp in position $i + 3$ and the backbone NH of the residue in position i) rather than from i to $i + 3$ backbone–backbone hydrogen bonds. This result is consistent with the observation[6] that residues with small polar sidechains, such as Ser, Thr, Asp and Asn, are found frequently in bends, presumably because these residues can interact most strongly with their immediate backbones.

From the above discussion, the following picture of the folding of the polypeptide chain emerges: helical (or other ordered structural) regions tend to form in certain regions of the amino acid sequence of the polypeptide chain, in response to short-range interactions. These are stabilized, however, only when long-range interactions come into play. This is brought about by the formation of β-turns among *specific* amino acid residues, also on the basis of short-range interactions, thereby enabling the ordered regions to approach each other. The remainder of the polypeptide chain then folds around these one or more regions of interacting ordered structures.

VIII. APPLICATION OF CONCEPT OF DOMINANCE OF SHORT-RANGE INTERACTIONS

It is possible that rough models of the three-dimensional structure of a protein can be formulated by applying the above ideas to predict the regions of the chain where helical regions and β-turns occur. It remains to be seen whether similar ideas can be used to predict where parallel and anti-parallel pleated sheet conformations arise. If such a rough model can be obtained, it should be possible to refine it by conventional energy-minimization procedures[1] without encountering the multiple-minima problem.

These ideas are presently being tested[31] by conformational energy calculations on trimers, pentamers, heptamers and nonamers from lysozyme. The conformation of the central residue of each of these oligopeptides is varied while that of the remainder of the peptide is maintained in the (x-ray determined) observed structure. Preliminary results indicate that the correct conformation of the central residue can be obtained (as that which minimizes the energy of the oligopeptide) when not only the short-range interactions in the central residue are included, but also longer and longer range interactions as the length of the oligopeptide is increased.

NOTE ADDED IN PROOF

1. *Table 1* has been extended to include L-phenylalanine[32] and L-valine[33].
2. It appears that extended (β) structures are also determined, in large measure, by short- and medium-range interactions[31, 34].

REFERENCES

[1] H. A. Scheraga, *Chem. Revs,* **71**, 195 (1971).
[2] D. Kotelchuck and H. A. Scheraga *Proc. Nat. Acad. Sci., Wash.* **61**, 1163 (1968).
[3] D. Kotelchuck and H. A. Scheraga, *Proc. Nat. Acad. Sci., Wash.* **62**, 14 (1969).
[4] A. V. Finkelstein and O. B. Ptitsyn, *J. Molec. Biol.* **62**, 613 (1971).
[5] P. J. Flory, *Principles of Polymer Chemistry*, chap. 14, Cornell University Press: Ithaca, N.Y. (1953).
[6] P. N. Lewis, F. A. Momany and H. A. Scheraga, *Proc. Nat. Acad. Sci., Wash.* **68**, 2293 (1971).
[7] A. V. Guzzo, *Biophys. J.* **5**, 809 (1965).
[8] B. H. Havsteen, *J. Theoret. Biol.* **10**, 1 (1966).
[9] J. W. Prothero, *Biophys. J.* **6**, 367 (1966).
[10] M. Schiffer and A. B. Edmundson, *Biophys. J.* **7**, 121 (1967).
[11] D. Kotelchuck, M. Dygert and H. A. Scheraga, *Proc. Nat. Acad. Sci., Wash.* **63**, 615 (1969).
[12] M. Gō, N. Gō and H. A. Scheraga, *J. Chem. Phys.* **54**, 4489 (1971).
[13] P. N. Lewis, F. A. Momany and H. A. Scheraga, *Israel J. Chem.* in press.
[14] B. H. Zimm and J. K. Bragg, *J. Chem. Phys.* **31**, 526 (1959).
[15] P. H. Von Dreele, D. Poland and H. A. Scheraga, *Macromolecules,* **4**, 396 (1971).
[16] P. H. Von Dreele, N. Lotan, V. S. Ananthanarayanan, R. H. Andreatta, D. Poland and H. A. Scheraga, *Macromolecules,* **4**, 408 (1971).
[17] V. S. Ananthanarayanan, R. H. Andreatta, D. Poland and H. A. Scheraga, *Macromolecules,* **4**, 417 (1971).
[18] K. E. B. Platzer, V. S. Ananthanarayanan, R. H. Andreatta and H. A. Scheraga, *Macromolecules,* **5**, 177 (1972).
[19] L. J. Hughes, R. H. Andreatta and H. A. Scheraga, *Macromolecules,* **5**, 187 (1972).
[20] J. Alter, G. T. Taylor and H. A. Scheraga, *Macromolecules,* **5**, 739 (1972).
[21] P. N. Lewis, N. Gō, M. Gō, D. Kotelchuck and H. A. Scheraga, *Proc. Nat. Acad. Sci., Wash.* **65**, 810 (1970).
[22] R. H. Pain and B. Robson, *Nature, London,* **227**, 62 (1970).
[23] P. N. Lewis and H. A. Scheraga, *Arch. Biochem. Biophys.* **144**, 576 (1971).
[24] N. Gō, P. N. Lewis, M. Gō, and H. A. Scheraga, *Macromolecules,* **4**, 692 (1971).
[25] P. N. Lewis and H. A. Scheraga, *Arch. Biochem. Biophys.* **144**, 584 (1971).
[26] K. Brew, T. C. Vanaman and R. L. Hill, *J. Biol. Chem.* **242**, 3747 (1967).
[27] R. L. Hill, K. Brew, T. C. Vanaman, I. P. Trayer and P. Mattock, *Brookhaven Symp. Biol.* **21**, 139 (1968).
[28] W. J. Browne, A. C. T. North, D. C. Phillips, K. Brew, T. C. Vanaman and R. L. Hill, *J. Molec. Biol.* **42**, 65 (1969).
[29] I. D. Kuntz, *J. Amer. Chem. Soc.* **94**, 4009 (1972).
[30] P. N. Lewis, F. A. Momany and H. A. Scheraga, *Biochim. et Biophys. Acta,* **303**, 211 (1973).
[31] P. K. Ponnuswamy, P. K. Warme and H. A. Sheraga, *Proc. Nat. Acad. Sci., Wash.* **70**, 830 (1973).
[32] H. E. Van Wart, G. T. Taylor and H. A. Scheraga, *Macromolecules,* **6**, 266 (1973).
[33] J. Alter, R. H. Andreatta, G. T. Taylor and H. A. Scheraga, *Macromolecules,* in press.
[34] P. K. Ponnuswamy, A. W. Burgess and H. A. Scheraga, submitted to *Israel J. Chem.*

ELECTRONIC–CONFORMATIONAL INTERACTIONS IN PROTEINS

M. V. Volkenstein

Institute of Molecular Biology, Academy of Sciences of the USSR, Moscow, USSR

ABSTRACT

The functionality of biopolymers and of the supermolecular structures in biology depends strongly on the interactions between electronic excitations and the shifts of the electronic density with the conformational changes in these systems. The study of the electronic–conformational interactions can help to form the physical theory of enzymatic activity and of the other biomolecular phenomena. The concept of the conformon can be introduced for the theoretical description of the electronic–conformational interactions, describing a kind of the quasi-particle containing the shift of the electronic density and the conformational changes. The direct experimental investigations of the electronic–conformational interactions were performed with the systems apo-aspartate-amino transferase with a series of different coenzymes. The denaturation of such systems (i.e. their conformational properties) depends strongly on the kind of the electronic interactions at the active site.

INTRODUCTION

The living organism and its functional systems including the molecules of proteins and nucleic acids are complex chemical machines characterized by the self-consistent regular and regulated behaviour of their elements. The main functions of biological systems are chemical, i.e. participation— primarily the catalytic one—in electronic transformations, and the realization of the electronic and ionic motions. In this sense the theoretical physical treatment of biomolecular systems and particularly of proteins has to be based on quantum mechanics.

The motion of the electrons in a molecule is coupled with the nuclear motion. The nuclear movements in biopolymers are specific—the lowest energy is required for the rotations of the atomic groups around single bonds, i.e. for conformational changes. The peculiarities of the structure and properties of the protein molecules are determined mainly by their conformational motility. The conformational and electronic motions of a protein molecule can be separated as the shifts of the nuclei are much slower than the electronic transitions. The cause of this separability is the same as in the theory of vibrational molecular spectra—the Born–Oppenheimer theorem. Therefore the physical theory of biopolymers, the theory of proteins, has to be based on investigation of the electronic–conformational interactions (ECI)[1]. The analysis of ECI must lead to an explanation of the main features of enzyme catalysis. On the other hand the behaviour of contractile

9

proteins responsible for biological mechanochemical processes such as muscular contraction, follows also from transconformations produced by some electrochemical phenomena. ECI build the fundamentals of mechano-chemistry.

In this paper we shall deal mainly with enzyme catalysis.

Koshland's theory, which has played a great role in the development of the modern theoretical treatment of enzymes, supports the great importance of transconformations for the electronic (i.e. chemical) processes catalysed by enzymes. However, this theory considers only the static pattern of the induced structural fit of enzyme and substrate but not the dynamical pheno-mena which determine the formation and further fate of the enzyme–substrate complex.

How can we approach the treatment of these dynamical phenomena on the basis of ECI?

Five years ago Perutz wrote that the enzyme acts not only as a reagent but also as a peculiar medium of reaction[2]. Undoubtedly, this is sound reasoning. The study of ECI needs the treatment of dynamical properties of such a medium possessing conformational motility. The theory of enzyme catalysis must be developed as a theory of chemical reactions occurring in specific condensed media characterized by conformational dynamics.

It is clear that the 'gaseous' theory of Eyring which takes into account the properties of a medium only via its macroscopic constant—the dielectric constant—is not sufficient for the dynamical treatment of ECI. In the works of Marcus[3] and in later more accurate studies of the problem[4, 5], the theory of an electronic exchange between ions in a condensed polar medium was developed. It is shown that all kinds of motions of particles taking part in the reaction have to be divided into classical (with small vibration frequencies $\omega \ll kT/\hbar$) and quantum-mechanical (with $\omega \gg kT/\hbar$) systems. The activation energy of the process is defined by the height of the barrier for the classical subsystem and the pre-exponential factor in the rate expression—by the tunnelling motion of the quantum subsystem below the barrier. The dynami-cal behaviour of the non-polar proteinic medium can be described in a harmonic approximation in terms of a set of acoustical vibrations, of the phonons. The probability of an elementary reaction act will be the sum of the products of the probability for the system to exist in a definite electronic and conformational state and the probability of transition of the system into another state. The consideration of both kinds of motions results in the change of the energetic map of the reaction compared with the purely electronic process. Here we have to look for the theoretical explanation of the complementarity of chemical and conformational energy suggested by Lumry and Biltonen[6].

The essential features of a non-polar protein medium are, first, the high entropy changes in the course of reaction. These are determined just by the pronounced acoustical nuclear motions. The second feature is that because of the small values of these acoustical frequencies the electronic subsystem follows the nuclear one adiabatically and the transmission coefficient is equal to unity[7].

We need of course not only the phenomenological general theory but also the theory of definite enzymatic processes taking into account the real transconformations and corresponding relaxation times (i.e. frequencies). These times can be determined with the help of the relaxational methods of chemical kinetics. The model of ECI which describes satisfactorily a lot of facts was suggested by Perutz for haemoglobin oxygenation[8].

On the basis of the general phenomenological description of ECI, a theory for the multicentrum catalysis was developed using the bridge mechanism[9]. The models of oxidative phosphorylation which are based on the same principles were proposed in other papers[10–12]. The quantum mechanical calculations of the electronic exchange in cytochrome, taking into account the dynamics of conformational motions, were recently performed by Madumarov[13].

The physical meaning of ECI is that the electronic transformation in the course of an enzymatic process determines the conformational changes in the macromolecule. Such a situation can be described in terms of modern solid state physics. The shift of an electron or of the electronic density in the macromolecule produces the deformation of the 'lattice', i.e. the conformational changes. It can be treated as the excitation of the long-wave phonons and the system electron plus conformational deformation of the macromolecule becomes like a polaron. Let us call such a system 'the conformon'[14].

The conformon is different from the polaron because the biopolymeric molecule lacks periodicity and homogeneity. Therefore the conformon is not a real quasi-particle. which can move far—its energy rapidly dissipates. However, for the realization of an enzymatic process the conformational change, i.e. the phonon excitation in the range of some peptidic bonds, is quite sufficient. The further transconformation is determined not by ECI but by the cooperative conformational movement of the nuclei in the macromolecule.

It seems natural to use the notion of the conformon for the study of the semi-conductive properties of biopolymers. However, the existence of such properties and their biological importance cannot be considered to be firmly established. Theory and experiment contradict the existence of semiconduction in proteins and nucleic acids. The observed conduction can be determined by some ionic contamination. It is difficult to imagine that non-coloured substances such as proteins and nucleic acids are electronic semiconductors.

On the other hand, ionic transport on biomembranes has to be studied, using the notion of the conformon. The physical theory of membranes can be built up with the help of the ideas from the modern physics of solids.

Let us now discuss the experimental study of ECI in the enzymatic processes.

EXPERIMENTAL STUDY OF ECI

One possible way to study ECI is the systematic investigation of changes in the chemical (electronic) properties of the protein produced by the change in the ligand or cofactor and of the simultaneous changes in the conformational behaviour of the macromolecule as a whole. The latter changes can express themselves in the course of denaturation and proteolysis. Such effects have been observed before in myoglobin[15]. The results obtained in

11

this work suggest that the changes in conformational properties of the protein as a whole can be produced by events occurring locally, at the active site. With the aim of studying ECI in detail we have investigated aspartate-amino transferase (AAT) with different modifications at the active site. This enzyme consists of two subunits with the molecular weight 47 000 each. The active site contains coenzyme—pyridoxal phosphate. The chemical transformations occurring on the active site of AAT during the action of enzyme were studied in detail by Braunstein and Shemyakin[16] and by Snell and co-workers[17]. Ivanov, Okina and other co-workers of our laboratory examined the denaturation in the urea solutions (pH 7.1, 0.05 M phosphate buffer) of the apoenzyme, normal holoenzyme, and of the complexes of the apoenzyme with modified coenzymes. Seven systems were studied as shown in *Figure 1*,

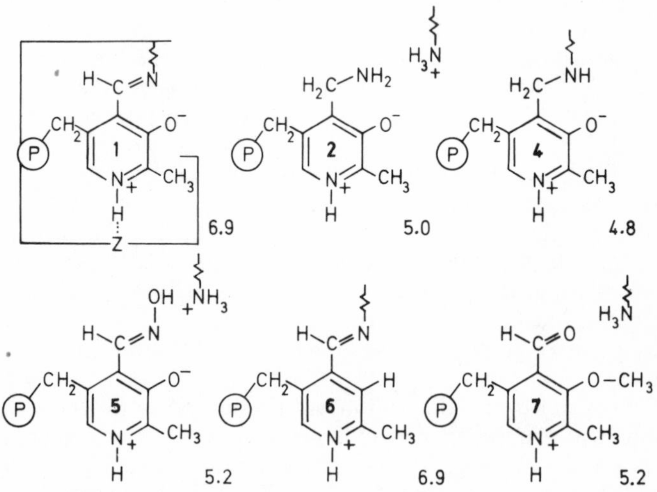

Figure 1. Modifications of the AAT active site.

where **1** is a holoenzyme with coenzyme in the aldimine form; **2** is a holoenzyme with coenzyme in the amino form (without covalent bond with ε-amino group of lysyl on the active site); **3** (not shown) is an apoenzyme, **4** is a holoenzyme treated with sodium borohydride ($NaBH_4$), i.e. with a reduced $C=N$ bond; **5** is a holoenzyme treated with hydroxylamine (NH_2OH), i.e. with a broken bond of the coenzyme with lysyl; **6** and **7** are complexes with modified coenzymes. *Figure 1* shows the values $C_m^\star$ of the urea concentration in moles, corresponding to the middle point of the denaturation isotherm. Denaturation was observed with the help of the CD band at 222 nm which characterizes the secondary structure. The equilibrium isotherms are shown in *Figures 2–4*. The seven systems studied correspond to two kinds of curves with $C_m^\star$ values near to 5.0 and 7.0. The systems whose electronic structures are very much alike, containing correspondingly the

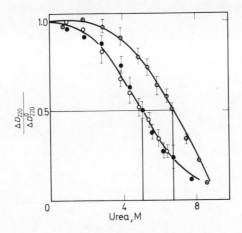

Figure 2. Denaturation curves for holo-AAR ($\oplus$), apo-AAT ($\bigcirc$) and holo-AAT + NH$_2$OH ($\bullet$).

C—O—CH$_3$ and C—H groups, give different curves. It has been shown earlier that the —O—CH$_3$ group hinders the formation of the aldimine bond with lysyl[18]. The curves for the amino form and for the holoenzyme treated with NaBH$_4$ are practically the same (*Figure 4*), notwithstanding that in one case (amino form) the aldimine bond does not exist and in another case this bond is very strong.

These results show directly that the electronic changes on the active site produce a strong effect on the conformational stability of the protein.

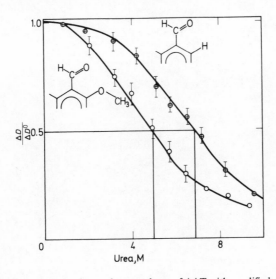

Figure 3. Denaturation isotherms for complexes of AAT with modified coenzymes.

13

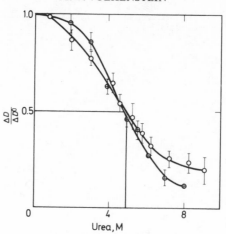

Figure 4. Denaturation isotherms for AAT treated with $NaBH_4$ ($\oplus$) and for AAT with coenzyme in amino form ($\bigcirc$).

The choice of AAT as the object for investigations of ECI was determined also by the multistage character of the transamination reaction catalysed by AAT. The stages of the reaction and their mechanisms have been described in detail[16, 17]. They are shown schematically in *Figure 5*. The numbers

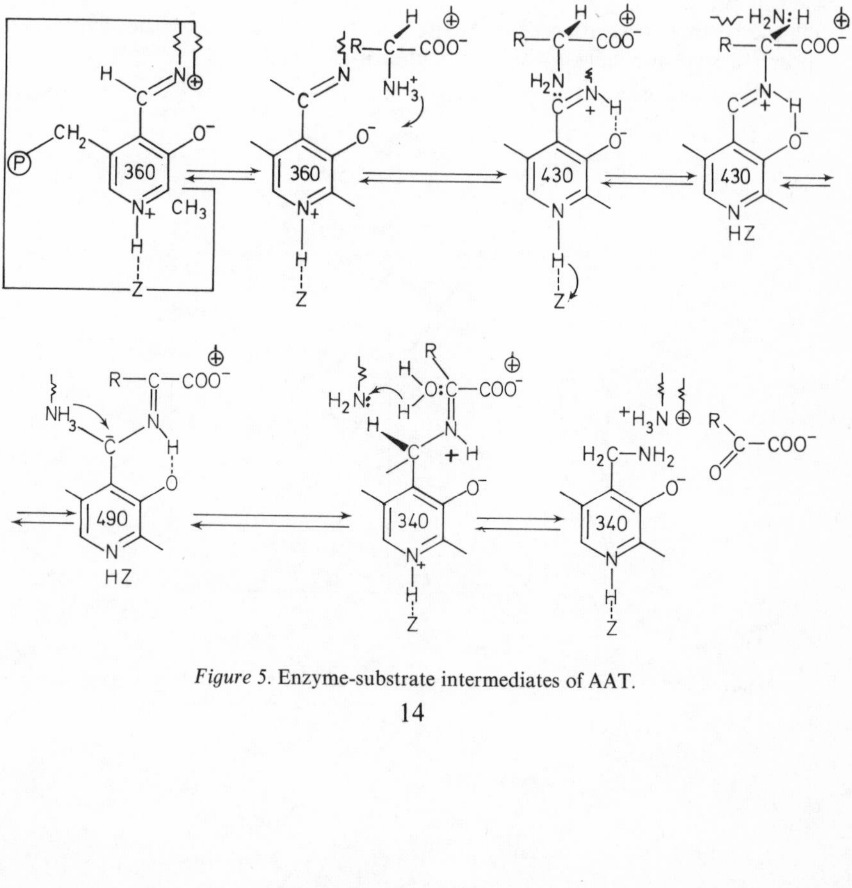

Figure 5. Enzyme-substrate intermediates of AAT.

14

in the rings give the wavelengths corresponding to the maxima of the absorption bands of these intermediate compounds.

An analogous reaction can occur in a congruent model system, containing a substrate, a free coenzyme and low molecular weight catalysts of proton separation and addition (i.e. imidazole). The reaction rate on such a system is lower by a factor $\sim 10^4$ than in an enzymatic one although the same inter-mediate chemical compounds were observed in the model system. In the model system the dominant forms correspond to absorption at 340 nm. This means that their free energy is lowest. In the enzymatic system all other forms are observed, including the form at 490 nm corresponding to the highest free energy. This means that the energetical levels of different forms do not differ much in the enzymatic system—they remain in the range of several kcal/mol. We see that the enzyme equalizes the free energy levels of different inter-mediate forms. Therefore the activation barriers between these forms become lower (Broensted's rule). According to the suggestion of Lumry and Biltonen[6] this equalization arises because of the complementarity of the curves of the chemical and conformational energy. It can be suggested that the changes of the conformational stability of the protein as a whole entity correlate with the free energy of the multistage process. If this is really so, then the conforma-tional stability of different forms arising in the multistage reaction has to be approximately complementary to the chemical free energies. We used the complexes of enzyme with inhibitors which stop the reaction at different stages as models of the intermediate forms. The conformational stability of these complexes was investigated. According to Tanford[19] from the values $C_m^\star$ the free energy of denaturation ΔF can be derived. *Figure 6* shows as

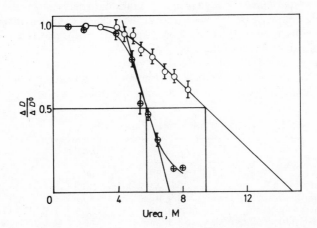

Figure 6. Denaturation isotherms of AAT + D,L-Asp ($\otimes$) and of AAT + glutarate ($\bigcirc$).

examples the denaturation isotherms for the complex of AAT with the substrate (corresponding to the equilibrium mixture of all intermediate forms) and for the complex of AAT with an inhibitor-glutarate modelling the sorption of the substrate. The curves differ considerably. In *Figure 7*, the values of ΔF are shown calculated from such curves according to Tanford for different complexes as models of the intermediate forms of the multistage reaction. We see that the conformational stability is different for different stages. The forms at 340 nm possessing the lowest chemical energy in the

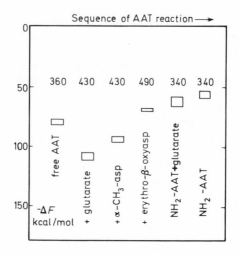

Figure 7. The free energies of denaturation of different AAT complexes.

congruent system have the highest conformational energy in correlation with the complementarity mentioned above. The exact complementarity is of course not observed as the complexes studied do not reproduce exactly the intermediate complexes of the reaction.

Let us summarize in *Table 1* all quantitative characteristics of denaturation —the $C_m^{\star}(M)$ values, the free energy changes and tan α values for the slope of the tangent of the denaturation curve at the point $C_m^{\star}$.

Table 1. Denaturation in urea solution

System	C(M)	F (kcal/mol)	tan α (M^{-1})
1. Holo-AAT	6.9 $\pm$ 0.2	79 $\pm$ 2	-1.8 ± 0.4
2. Aminoform of AAT	5.0 $\pm$ 0.3	58 $\pm$ 3	-1.3 ± 0.3
3. Apo-AAT	5.2 $\pm$ 0.3	60 $\pm$ 3	-1.6 ± 0.4
4. AAT reduced by NaBH$_4$	4.8 $\pm$ 0.4	54 $\pm$ 4	-1.8 ± 0.4
5. AAT + NH$_2$OH	5.2 $\pm$ 0.3	60 $\pm$ 3	-2.0 ± 0.9
6. Apo-AAT + 3-deoxy-PLP	6.9 $\pm$ 0.3	80 $\pm$ 3	-1.3 ± 0.3
7. Apo-AAT + 3-methoxy-PLP	5.1 $\pm$ 0.2	59 $\pm$ 2	-1.4 ± 0.4
8. AAT + D,L-aspartate	5.8 $\pm$ 0.3	67 $\pm$ 3	-3.5 ± 1.1
9. AAT + α-methyl-aspartate	8.1 $\pm$ 0.2	94 $\pm$ 2	-1.9 ± 0.5
10. AAT + erythro-β-oxyaspartate	6.1 $\pm$ 0.1	70 $\pm$ 1	-2.8 ± 1.0
11. AAT + glutarate	9.4 $\pm$ 0.3	108 $\pm$ 3	-1.0 ± 0.2
12. Amino-AAT + glutarate	5.6 $\pm$ 0.4	64 $\pm$ 4	-2.0 ± 0.6

16

Similar information concerning ECI can be obtained from the studies of proteolysis. Kinetics and thermodynamics of the proteolytic splitting of a protein correlates with its denaturation behaviour and characterizes the conformational properties of the whole molecule. We have obtained some preliminary results corresponding to this suggestion.

In this way we obtained some early results in our theoretical and experimental investigations of ECI. These phenomena are extremely important as they determine the physicochemical and therefore the biological properties of biopolymers.

REFERENCES

[1] M. Volkenstein. *Izvest. Akad. Nauk. SSSR, Ser. Biol.* No. 6, 805 (1971).

[2] M. Perutz. *Proc. Roy. Soc. B*, **167**, 448 (1967).

[3] R. Marcus. *J. Chem. Phys.* **24**, 966 (1956); **26**, 867 (1957).

[4] R. Dogonadze and A. Kusnetzov. *Elektrochimia*, **3**, 1324 (1967);
R. Dogonadze and A. Kusnetzov. *Itogi Nauki. Elektrochimia* 1967. VINITI: Moscow (1969).

[5] R. Dogonadze, A. Kusnetzov and V. Levitch. *Dokl. Akad. Nauk SSSR*, **188**, 383 (1969).

[6] R. Lumry and R. Biltonen in *Structure and Stability of Biological Macromolecules*. Marcel Dekker: New York (1969).

[7] M. Volkenstein, R. Dogonadze, A. Madumarov, Z. Urushadze and J. Harkatz. *Molec. Biol.* **6**, 431 (1972).

[8] M. Perutz. *Nature, London*, **228**, 726 (1970).

[9] A. Madumarov. *Thesis*. Institute of Chemical Physics of the Academy of Sciences of the USSR.

[10] L. Blumenfeld and V. Koltover. *Molec. Biol.* **6**, 161 (1972).

[11] N. Chernavskaja, D. Chernavsky and Grigorow in *Vibrational Processes in Biological and Chemical Systems*, Vol. II, p 72 Pushtshino: Nauka (1971) (in Russian).

[12] D. Green. *Proc. Nat. Acad. Sci., Wash.* **67**, 544 (1970).

[13] A. Madumarov. *Molec. Biol.* in press.

[14] M. Volkenstein. *J. Theor. Biol.* **34**, 193 (1972).

[15] B. Atanasov, J. Sharonov, A. Shemelin and M. Volkenstein. *Molec. Biol.* **1**, 477 (1967).

[16] A. Braunstein and M. Shemyakin. *Biochimia*, **18**, 393 (1953).

[17] D. Metzler, M. Ikawa and E. Snell. *J. Amer. Chem. Soc.* **26**, 648 (1954).

[18] S. Mora, A. Bocharov, V. Ivanov, M. Karpeisky, O. Mamaeva and N. Stambolieva. *Molec. Biol.* **6**, 119 (1972).

[19] C. Tanford. *J. Amer. Chem. Soc.* **86**, 2050 (1964).

SOME ASPECTS OF NMR TECHNIQUES FOR THE CONFORMATIONAL ANALYSIS OF PEPTIDES

V. F. BYSTROV*, S. L. PORTNOVA, T. A. BALASHOVA, S. A. KOZ'MIN, YU. D. GAVRILOV and V. A. AFANAS'EV

Shemyakin Institute for Chemistry of Natural Products, USSR Academy of Sciences, Moscow, USSR

ABSTRACT

An outline is given of some new NMR spectroscopic approaches to the spatial structure of peptides in solution which have been recently suggested and tested in the Shemyakin Institute for Chemistry of Natural Products. The previously derived angular dependence of the peptide vicinal $^3J_{NHCH}$ coupling constant has been refined on the basis of the latest experimental data. It has been found that ion–dipole interaction of the type $C{=}O \cdots M^+$ (where M^+ is an alkali metal ion) leads to a low field shift of the ^{13}C signal of the carbonyl group. This effect permits determination of the number and location of the ligand groups which form the internal cavity of peptide and depsipeptide complexones. The effect of 'shift reagents' on the NMR spectra sheds certain light on the spatial structure of peptides in solution, and, in particular, gives considerable information on the configuration of the amide bond and on the rotational states of the C^α—C^β bond. The INDOR and signals on combination frequency have been used for detection of 'hidden' signals (their multiplicity, chemical shift and splitting) and on the assignment of the NH signals in the proton NMR spectra of peptides.

High resolution NMR spectroscopy is becoming the most powerful means for the study of the spatial structure of peptides in solution. The progress in this field has required the development of new spectroscopic approaches. The subject matter of this paper is to outline some of those which have been suggested and tested recently in the Shemyakin Institute for Chemistry of Natural Products.

I. THE ANGULAR DEPENDENCE OF THE PEPTIDE VICINAL NH-C$^\alpha$H COUPLING CONSTANT

With the accumulation of more experimental data it has now become possible to refine the earlier proposed[1,2] dependence of the $^3J_{NHCH}$ constant on the dihedral angle θ between the H—N—C$^\alpha$ and N—C$^\alpha$—H planes.

We assume as before[1] that our dependence is expected in the form of a Karplus-like equation[3]

$$^3J_{\text{NHCH}} = A \cos^2\theta - B \cos\theta + C \sin^2\theta \qquad (1)$$

where A, B and C are positive coefficients.

The corrections for the C^α substituent electronegativity are made according to

$$^3J_{\text{NHCH}} = J_{\text{obs}}\left(1 - \alpha \sum_i \Delta E_i\right)^{-1} \qquad (2)$$

where J_{obs} is the experimental value for the NH—C^αH coupling, and ΔE_i is the electronegativity difference between the C^α substituents and hydrogen. Assuming the value $\alpha = -0.1$ in conformity with the experimental data for ethanes[4] and the Pauling scale of electronegativity[5], equation 2 for the peptide fragment

$$\text{NH—C}^\alpha\text{H} \overset{\displaystyle C^\beta}{\underset{\displaystyle C'}{\diagdown}} \qquad \text{becomes} \quad J_{\text{NHCH}} = 1.09 \, J_{\text{obs}} \qquad (3)$$

In what follows only constants corrected according to equation 3 are used.

For any torsional potential of internal rotation as well as for free rotation about the N—CH_3 bond in the $\overset{O}{\underset{}{\diagdown}}C\text{—N}\overset{CH_3}{\underset{H}{\diagup}}$ fragment the averaged coupling constant is

$$^3J_{\text{NHCH}_3} = (A + C)/2$$

The experimental data are 4.9 Hz for N-methylformamide[6,7] and 4.8 Hz for N-methylacetamide[8].

In addition new maximum $^3J_{\text{NHCH}}$ values for the regions $0° \leqslant \theta \leqslant 90°$ and $90° \leqslant \theta \leqslant 180°$ were selected from the data on peptides with *trans*-amide bonds. For the first region the value of 8.0 Hz was selected, being the constant of the NH—C^αH fragment of the D-Val residue of valinomycin in the 'bracelet' conformation in non-polar media[9]. For the second region the maximum constant of 10.2 Hz was selected as that exhibited by the corresponding fragments of the L-Val and L-Leu residues of gramicidin S[10].

We thus have the following boundary conditions for the coefficients of equation 1:

$$A + C = 9.8 \text{ Hz}$$

$$A - B \geqslant 8.0 \text{ Hz}$$

$$A + B \geqslant 10.2 \text{ Hz}$$

With these constraints computer calculation gave the angular dependence of the $^3J_{\text{NHCH}}$ coupling constant presented in *Figure 1*. The refined dependence lies within the region of the one proposed earlier[1,2], but confines permissible angles for a given experimental coupling constant to narrower limits and more definitely discriminates between the *cis*- and *trans*-orientation of the NH—C^αH protons.

The mean permissible values shown in *Figure 1* as a hatched area are approximated by equation

$$^3J_{\text{NHCH}} = 9.8 \cos^2\theta - 1.1 \cos\theta + 0.4 \sin^2\theta$$

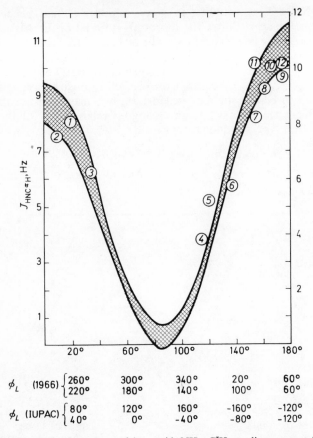

| ϕ_L | (1966) | $\begin{cases} 260° \\ 220° \end{cases}$ | $\begin{matrix} 300° \\ 180° \end{matrix}$ | $\begin{matrix} 340° \\ 140° \end{matrix}$ | $\begin{matrix} 20° \\ 100° \end{matrix}$ | $\begin{matrix} 60° \\ 60° \end{matrix}$ |
| ϕ_L | (IUPAC) | $\begin{cases} 80° \\ 40° \end{cases}$ | $\begin{matrix} 120° \\ 0° \end{matrix}$ | $\begin{matrix} 160° \\ -40° \end{matrix}$ | $\begin{matrix} -160° \\ -80° \end{matrix}$ | $\begin{matrix} -120° \\ -120° \end{matrix}$ |

Figure 1. Refined angular dependence of the peptide NH—C$^\alpha$H coupling constant. Experimental data: *1* and *3*, D- and L-Val residues of valinomycin in the 'bracelet' conformation[9]; *2* and *10*, L- and D-Val of valinomycin in the 'propeller' conformation[9, 11]; *4, 9, 11* and *12*, D-Phe, L-Orn, L-Leu and L-Val of gramicidine S[10]; *5*, L- and D-Val of K$^+$-valinomycin complex[9, 11]; *6–8*, L-Orn$_{2, 3, 1}$ of alumichrome A[12] (x-ray diffraction data for ferrichrome A, see ref. 13).

which on rearrangement gives

$$^3J_{\text{NHCH}} = 9.4 \cos^2\theta - 1.1 \cos \theta + 0.4$$

Figure 1 also presents the most reliable experimental data obtained by the composite physicochemical method of conformational analysis in solution[9–11] and by x-ray analysis for peptides with *trans*-amide bonds[12, 13]. The extreme values for $^3J_{\text{NHCH}}$ known up to now (2.7 Hz for evolidine[14]; less than 2.6 Hz for the Na$^+$-antamanide complex[15] and the maximum value of 11.7 Hz for one of the forms of 'symmetric' Val$^{(6)}$, Ala$^{(9)}$-antamanide[16]) are in good accord with the general range of values for this newly derived dependence (*Figure 1*).

21

Vicinal NH—$C^{\alpha}H_2$ coupling in glycyl residues should be considered separately as the spectrum of these protons is either of the ABX or AA'X type. In this case the line separation of the NH signal (X-proton)—quartet or triplet—as a rule does not directly give the $^3J_{NHCH}$ coupling constants[17]. Only the separation between the outer components is strictly equal to the sum of J_{AX} and $J_{BX}(\Sigma^3J_{NHCH_2})$. Assuming the projection angle between the $NC^{\alpha}H$ glycyl planes to be the standard value 120°, one may then use the above $^3J_{NHCH}$ dependence to calculate an analogous relationship for the overall constant of the glycyl residue protons from

$$\Sigma^3J_{NHCH_2} = {}^3J_{NHCH}(\theta) + {}^3J_{NHCH}(120° \pm \theta)$$

The result obtained is shown in *Figure 2* as a function of the conventional

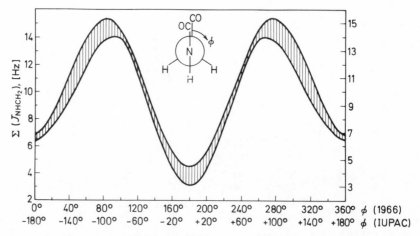

Figure 2. Overall vicinal coupling constant in the glycyl NH—$C^{\alpha}H_2$ fragment plotted against the conformational angle ϕ[18, 19].

conformational angles ϕ[18, 19]. The mean permissible values are approximated by the following function of the IUPAC–IUB angle ϕ[18]

$$\Sigma^3J_{NHCH_2} = -9.8 \cos^2\phi - 1.3 \cos\phi + 15.0$$

The experimental glycyl coupling constants ranging from 13.5 Hz* for glycylalanyl cyclopeptides[20] to 7.7 Hz for alumichrome[12] fall within the extremes of the $\Sigma^3J_{NHCH_2}$ curve (*Figure 2*).

Recently several other sets of coefficients for equation 1 have been proposed[21–24] (*Table 1*). The coefficients suggested by Schwyzer[23] practically coincide with those for a curve passing through the average values of our band dependence for the free rotation model[2]. The theoretically calculated dependence of Karplus and Barfield[21] does not reflect the actual difference between the *cis*- and *trans*-oriented NH—$C^{\alpha}H$ protons. The empirical

---

*Electronegativity corrections given according to expression $\Sigma^3J_{NHCH_2} = 1.04\,\Sigma J_{obs}$.

dependences of Thong et al.[22] and of Ramachandran et al.[24] give $^3J_{NHCH}$ values too low for the cis-oriented NH and C^αH bonds ($\theta = 0°$). This must be due to the fact that these authors used six-membered heterocycles with cis-amide bonds for defining the $0° \leqslant \theta \leqslant 90°$ region of the curve. Quite possibly the strain occurring in these compounds, particularly in iso-quinuclidone (as shown in ref. 24), leads to somewhat smaller $^3J_{NHCH}$ constants than would have been found for the trans-amide bonds. The pre-liminary theoretical INDO calculation shows that indeed the NH—C^αH coupling constant is smaller for compounds with a cis-amide bond than with a trans-amide bond[25].

Table 1. Coefficients proposed for equation 1.

Reference	Coefficients in Hz			$^3J_{NHCH}$ in Hz	
				Cis ($\theta = 0°$)	Trans ($\theta = 180°$)
	A	B	C		
This paper[a]	9.8	−1.1	0.4	8.0–9.4	10.2–11.6
Bystrov et al.[1,a]	9.3	−0.5	0.9	8.0–9.8	8.0–11.6
Bystrov et al.[2,a]	9.6	−0.2	0.4	8.9–9.8	8.9–10.7
Barfield and Karplus[21]	12.0	0.0	0.2	12.0	12.0
Thong et al.[22,b]	9.3	−3.5	0.3	5.8	12.8
Schwyzer[23]	9.68	−0.42	0.12	9.26	10.10
Ramachandran et al.[24,b]	8.6	−1.7	1.5	6.9	10.3

[a]The coefficients correspond to a curve in the middle of the allowed region. [b]Electronegativity corrections made according to equation 3.

The good agreement between the experimental data and the derived curves gives ground to believe that the proposed angular dependence will provide a more precise conformational coordinate and will find ever increasing use in studies of peptides by proton NMR. Dihedral angles θ can be converted into the conventional conformational angles ϕ[18, 19] as shown in Table 2.

II. DETECTION OF ION–DIPOLE INTERACTIONS IN PEPTIDE COMPLEXES WITH ALKALI METAL BY ^{13}C NMR

It is to be expected that the intensively developing field of ^{13}C NMR will open up new possibilities in studies of the spatial structure of peptides. As a first step in this direction we reported[26] on the effect on the ^{13}C spectra of the formation by cyclic depsipeptides and peptides of stable alkali ion complexes (for the independent communication on this effect see ref. 27).

It is known that the distinctive characteristic of the peptidic complexones (see, for example, refs. 9, 11, 15, 16) is the location of the ion within the central cavity of the cyclic molecule so that the former is held in place by ion–dipole interaction with those carbonyls that are oriented towards the centre of the cavity. Such compounds are being widely used as tools for the study of processes associated with ion transport through membranes. Among the most popular of the alkali ion complexones are the depsipeptide antibiotics valinomycin and the enniatins, and the cyclic decapeptide antamanide.

Table 2. Relation between the dihedral angles θ and the conventional

		θ	0°	20°	40°
According to the nomenclature (1966)[18]					
	ϕ	L-	240°	260° 220°	280° 200°
		D-	120°	140° 100°	160° 80°
According to the IUPAC nomenclature[19]					
	ϕ	L-	60°	80° 40°	100° 20°
		D-	−60	−40° −80°	−20° −100°

From general considerations it is to be expected that with ion–dipole interaction of the type $C=O \cdots M^+$, where M^+ stands for the monovalent cation (Na^+ or K^+), an additional shift of electron density on the carbonyl bond toward the oxygen will take place. This should decrease the ^{13}C screening only of those carbonyls of the peptidic complexone that are sufficiently near to the cation.

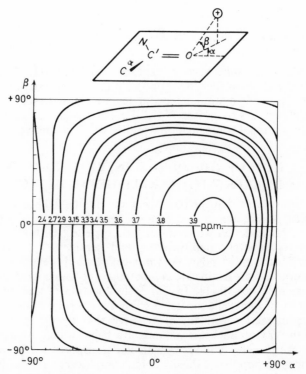

Figure 3. Influence of the positive charge on the ^{13}C screening of carbonyl as a function of the angles α and β for the $O \cdots M^+$ distance 2.8 Å.

angles ϕ for L- and D-amino acid residues in peptides.

60°	80°	100°	120°	140°	160°	180°
300°	320°	340°	360°,0°	20°	40°	60°
180°	160°	140°	120°	100°	80°	
180°	200°	220°	240°	260°	280°	300°
60°	40°	20°	0° 360°	340°	320°	
120°	140°	160°	±180°	−160°	−140°	−120°
0°	−20°	−40°	−60°	−80°	−100°	
0°	20°	40°	60°	80°	100°	120°
−120°	−140°	−160°	±180°	160°	140°	

The rough estimation of the expected ^{13}C shift of the carbonyl has been carried out by the theory of magnetic screening associated with chemical bond polarization under the influence of a point electrostatic charge[28, 29]. Taking into account only the first order effects one may express the change of ^{13}C screening for the amide bond $\overset{N}{\underset{C^\alpha}{\diagdown}} C'{=}O \cdots M^+$ as

$$\Sigma\Delta\sigma = -A_{C'=O}E_{C'=O} - A_{C'-N}E_{C'-N} - A_{C'-C^\alpha}E_{C'-C^\alpha}$$

where A denotes the coefficients depending on the bond polarizability; and E, the components of the electrostatic field of the point charge M^+ along with the direction of the corresponding chemical bond:

$$E_{C'-X} = |e|r^{-2}\cos\chi_{X-C'}$$

Here r is the distance from the positive charge to the carbonyl carbon atom, and $\chi_{X-C'}$ is the angle between the direction $M^+ \cdots C'$ and the X—C′ bond). The screening for the ester fragment $\overset{O}{\underset{C^\alpha}{\diagdown}} C'{=}O \cdots M^+$ is expressed in a similar manner.

Adopting the A values as in refs. 29, 30 one could calculate the total shift $\Sigma\Delta\sigma$ as a function of the angular coordinates α and β of the M^+ ion with regard to the carbonyl bond (*Figure 3*). As the result we have found that (1) the ^{13}C signal must undergo the low field shift; (2) the total shift has the maximum value −3.9 p.p.m. (for $r = 2.8$ Å) and depends on the α and β angles; (3) the $\Sigma\Delta\sigma$ values and the general appearance of their dependence on the angles are similar for the amide and ester groups.

These conclusions are confirmed by the following experimental data.

With valinomycin† $\ulcorner$(D-Val—L-Lac—L-Val—D-HyIv)$_3$$\urcorner$ we have the following: there are four $^{13}C{=}O$ signals (see *Table 3*) of which, owing to the symmetry of the chemical and spatial structure of the molecule[9, 11], each should correspond to the carbonyls of three identical amino or hydroxy acid

---

†The following abbreviations for residues are used: HyIlo, α-hydroxyisovaleric acid; Lac, lactic acid; MePhe, N-methylphenylalanine; MeLeu, N-methylleucine; MeIle, N-methylisoleucine.

Table 3. Chemical shifts (± 0.2 p.p.m.) of $^{13}C{=}O$ signals of depsipeptidic and peptidic complexones (with reference to carbon disulphide as internal standard)[a].

Valinomycin		Beauvericin			"Symmetric" Val[(6)], Ala[(9)]-antamanide	
Non-complexed	K^+ complex	Non-complexed	K^+ complex	Na^+ complex	Non-complexed	Na^+ complex
20.4 (ester)	16.8	21.6	20.2	19.9	19.0	16.7
20.9 (amide)	19.8	22.9	20.9	20.6	20.8	19.3
21.7 (ester)	17.3				21.5	21.0
22.3 (amide)	21.2				21.5	21.7
					21.8	22.1

[a]The ^{13}C spectra were obtained at 22.63 MHz on a Bruker HX 90/18—18″ spectrometer in the FT mode (2 000 to 8 000 scans at 0.8 s/scan). Solutions of 100 to 300 mg of substance in 1.4 ml of a 1:1 (v.v) mixture of $CDCl_3$ with CD_3OD were used. The complexes were formed by adding KNCS or NaNCS in 1:2.5 to 1:5 mole ratios to the compound.

residues. Now in the spectrum of the K^+ complex only one $^{13}C{=}O$ signal falls within the $C{=}O$ region of the non-complexed valinomycin spectrum. According to the assignment given by Ohnishi *et al.*[27] the two ester $^{13}C{=}O$ signals undergo a larger downfield shift (-3.6 and -4.4 p.p.m.) than the two amide $^{13}C{=}O$ signals (both -1.1 p.p.m.). This is in complete agreement with the spatial structure of the K^+–valinomycin complex [*Figure* 4(a)] of

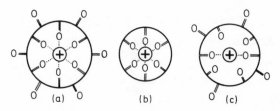

Figure 4. Schematic representation of the carbonyl conformations in alkali ion complexes of valinomycin (a), enniatin B and beauvericin (b), and antamanide (c). For detailed conformation of the compounds see refs. 9, 31, 15, respectively.

which only ester carbonyls are engaged in strong ion–dipole interaction with the K^+ ion located in the molecular cavity (the $O \cdots K^+$ distance is 2.7 to 2.9 Å). The direction and the values of these carbonyl shifts are consistent with the above rough calculations. The noticeable paramagnetic shift of the amide $^{13}C{=}O$ signals seems to mean that, besides taking part in the intramolecular hydrogen bonding, these carbonyls may also be participating in a weak ion–dipole interaction with the K^+ ion (here the $O \cdots K^+$ distance is around 4.0 to 4.5 Å).

In conformity with the theory of magnetic screening associated with electro-static bond polarization it follows that the induced chemical shift should decrease with decrease in the $C=O\cdots M^+$ angle from $180°$ to $90°$. In cyclohexadepsipeptidic enniatin complexes where all six carbonyls are symmetrically located around the central cation[31] [*Figure 4(b)*], these angles are much less than $180°$ (in contrast with valinomycin where the corre-sponding angles are approximately $180°$[9, 11]). Hence, despite the practically identical $O\cdots K^+$ distance in both types of complexes, the change in $^{13}C=O$ resonance should be less for enniatins. In fact, in the K^+ complex of beau-vericin† $\overline{(L\text{-MePhe}—D\text{-HyIv})_3}$ the $^{13}C=O$ signals undergo less change in position than in the valinomycin complex (*Table 3*).

When the K^+ cation is replaced by the smaller Na^+ cation, there is a characteristic change in the enniatin conformation, similar to the closing of a flower[31]. This is connected with a decrease in the $O\cdots M^+$ distance and a straightening out of the $C=O\cdots M^+$ angle. Both of these spatial structure changes should cause an increase in the downfield shift of the corresponding $^{13}C=O$ signals. Indeed, the $^{13}C=O$ resonances for the Na^+ complex of beauvericin are in lower field than those for the K^+ complex (*Table 3*).

In the ^{13}C spectrum of the Na^+ complex of a symmetrical analogue of antamanide –Val$^{(6)}$,Ala$^{(9)}$–antamanide:

$$\overline{(\text{L-Val—L-Pro—L-Pro—L-Ala—L-Phe})}$$
$$1,6 \quad\quad 2,7 \quad\quad 3,8 \quad\quad 4,9 \quad\quad 5,10$$

a significant shift to lower field (not less than -2.3 p.p.m.) is observed for only one $^{13}C=O$ line corresponding to two symmetrically located identical amino acid residues (*Table 3*). The shifts of the other four $^{13}C=O$ signals are within the limits of only $+0.5$ to -1.5 p.p.m. Hence the carbonyls of only two identical residues (apparently Val$^{(1)}$ and Val$^{(6)}$) of the cyclodecapeptide approach the cation located within the internal cavity. This conclusion is in complete agreement with the earlier proposed conformation of the Na^+–antamanide complex[15] of which only two carbonyls of the symmetric-ally situated Val$^{(1)}$ and Phe$^{(6)}$ residues are in close contact (*ca.* 2.6 Å) with the central ion [*Figure 4(c)*].

Besides the ion–dipole interaction induced shifts of the $^{13}C=O$ signal discussed above, marked changes are observed also in the positions of a number of other signals[26]. For instance, both C^α signals from D- and L-Val residues of valinomycin are shifted by 2.1 to 3.4 p.p.m. to the low field on complex formation, whereas the C^α signals from L-Lac and D-HyIv residues are shifted by only -1.2 and -0.8 p.p.m. Obviously such selective shifts are due to a conformational rearrangement of the molecule on complex formation[9, 11].

Undoubtedly ^{13}C NMR will become a powerful new method among the physicochemical techniques used for investigating the secondary and tertiary structures of peptide systems. A particular stress is to be placed on the considerably higher 'resolving power' of ^{13}C spectroscopy compared with proton NMR. For instance, in the ^{13}C spectra of the most complicated

†Beauvericin stands very close in conformational parameters and complexing properties to the thoroughly investigated[31] enniatin B $\overline{(L\text{-MeVal—D-HyIv})_3}$.

27

compound in this study, 'symmetric' antamanide and its Na$^+$ complex, individual signals are clearly visible from all of the 25 structurally non-equivalent carbon atoms. Evidently, the main difficulty that can be seen so far is the assignment of the ^{13}C signals, in particular those of the carbonyl groups. This can, of course, be overcome by synthesis of compounds selectively labelled with the ^{13}C or ^{15}N isotope.

III. THE APPLICATION OF 'SHIFT REAGENTS' IN THE NMR SPECTROSCOPY OF PEPTIDES†

Another intensively developing field of NMR spectroscopy is the use of the so-called 'shift reagents', a rapidly expanding class of substances, mostly lanthanide complexes such as tris(dipivaloylmetanato)europium(III) [Eu-(DPM)$_3$] as aids in spectral and structural interpretations[32]. Their name stems from the ability of these reagents to associate with polar functional groups and thereby cause substantial specific shifts of the NMR signals of the closely lying nuclei, the magnitude of the shift being largely determined by the distance of the nucleus from the lanthanide ion.

In peptides, such polar groups would be the carbonyls and one could have expected that the 'shift reagents' could shed some light on the spatial arrangement of magnetic nuclei, mainly protons and ^{13}C surrounding these groups. In particular, we have made use of the shift reagents for determining the configuration of the amide bond and of the rotational states of the C$^\alpha$–C$^\beta$ fragments.

The assignment of NMR signals to isomers with *cis*- and *trans*-amide bonds is a matter of some difficulty. Apparently it is only the 'shift reagents' which permit a general solution of this problem. It has been shown[33] that with the simple *N*-alkylamides the 'shift reagent' moves the signal of the N—CH$_3$ group *cis* to the carbonyl oxygen much farther downfield (-9.3 p.p.m.) than the *trans*-N—CH$_3$ signal (-4.0 p.p.m.)‡. Obviously this is due to the *cis*-methyl group being located considerably nearer to the complexing position than the *trans*-methyl group. We have obtained similar results for some model and biologically active peptides and depsipeptides with *N*-methylated amide bonds.

One of the simplest linear peptides, dipeptide *N*-benzyloxycarbonyl-D-alanyl-L-*N*-methylalanine methyl ester (I) exists as a 3:7 mixture of *cis*- and *trans*-isomers with respect to the *N*-methylamide bond[1, 34]:

cis   trans

---

†This part has been done in collaboration with Dr V. P. Zvolinskii.
‡ The shifts of the signals are extrapolated to 1:1 mole 'reagent'—compound ratio.

As expected, the addition of Eu(DPM)$_3$ shifts the N—CH$_3$ signal of the *cis*-isomer much more (-9.1 p.p.m.) than that of the *trans*-isomer (-2.2 p.p.m.)†. This then is a direct and convenient method for assigning signals to configurational isomers of *N*-alkylated peptides.

It is noteworthy that the ester CH$_3$ signal undergoes a rather small shift: -0.4 and -1.6 p.p.m. for *trans*- and *cis*-isomers of the dipeptide (I), respectively. Apparently the ester group forms a less stable complex with Eu(DPM)$_3$ than does amide carbonyl[35]. This conclusion is important for the NMR study of depsipeptides.

The diketomorpholines [for example (II) to (IV)], the simplest cyclic depsipeptides, represent convenient model compounds with *cis*-amide bonds. The Eu(DPM)$_3$ reagent

DD (II) and DL (III)   LD (IV)

caused a downfield shift of the N—CH$_3$ signals in accord with the values cited above for *cis*-NCH$_3$ groups: (II) -9.3 p.p.m., (III) -9.2 p.p.m. and (IV) -9.5 p.p.m.

The cyclotetradepsipeptide (V) contains two amide bonds

$$\overline{\text{L-MeIle—D-HyIv—L-MeLeu—D-HyIv}}\rfloor \qquad\qquad (V)$$

and the observed shifts when extrapolated to the ratio of one mole Eu(DPM)$_3$ per one amide bond were found to be -10.2 and -9.9 p.p.m. indicative of *cis*-configuration, both amide bonds showing them to be of the same configuration in a solution of the depsipeptide (V) as determined by x-ray analysis[36] for the solid state.

The *trans*-configuration of the amide bonds in biologically active cyclohexadepsipeptide enniatin B[31] (VI) was confirmed by the relatively low Eu(DPM)$_3$ induced shift of

$$\overline{\text{(L-MeVal—D-HyIv)}_3}\rfloor \qquad\qquad (VI)$$

N—CH$_3$ signal: -2.0 p.p.m. per one amide bond. This value closely agrees with the above shift for the *trans*-isomer of (I).

The good correlation of the 'shift reagent' induced shifts for the cyclic depsipeptides (II) to (VI) on the one hand and for dipeptide (I) and the *N*-alkylamides[33] on the other hand confirms the above observation that the ester groups associate more weakly with Eu(DPM)$_3$ than do the amide groups.

In addition to the method discussed above for determining the amide bond configuration, one might expect that the 'shift reagents' could be used

---

†The ^{1}H NMR spectra were obtained for 0.1 to 0.2 mol/l. solutions in CDCl$_3$ with increasing amounts of Eu(DPM)$_3$ up to 0.4 mole ratio of 'shift reagent' to compound.

29

as a supplementary tool for determining the rotational states of the C^α—C^β bond in the amino and hydroxy acid residues. From the NMR spectra it is as yet impossible to define which of the rotamers b or c (*Figure 5*) is predominant. However, the 'shift reagent' on associating with the carbonyl should be

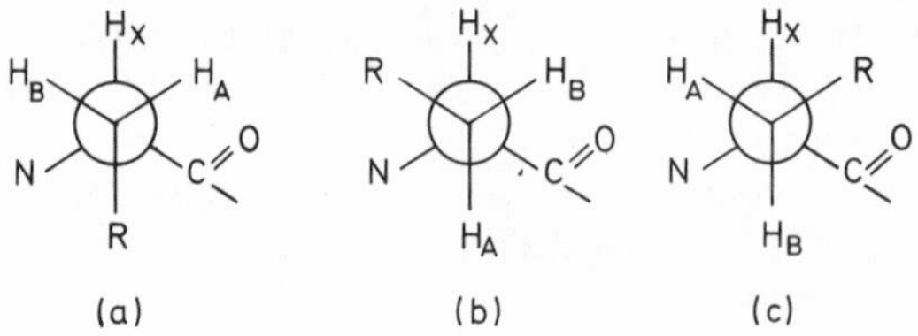

(a)   (b)   (c)

Figure 5. Rotational states of the $C^\alpha H_X$—$C^\beta H_A H_B$ fragment in the amino or hydroxy acid sidechain.

expected to cause an equal and considerable shift of both magnetically non-equivalent H_A and H_B protons in rotamer b but considerably differing shifts for these protons in rotamer c.

This is confirmed by the following example. The sidechain of the HyIv residue of the diketomorpholines (II) to (IV) could assume the rotational states shown in *Figure 6*. The values of the vicinal ^3J ($C^\alpha H_X C^\beta H_A$) coupling constant for compounds (III) and (IV) (2.0 and 2.3 Hz, respectively) indicate

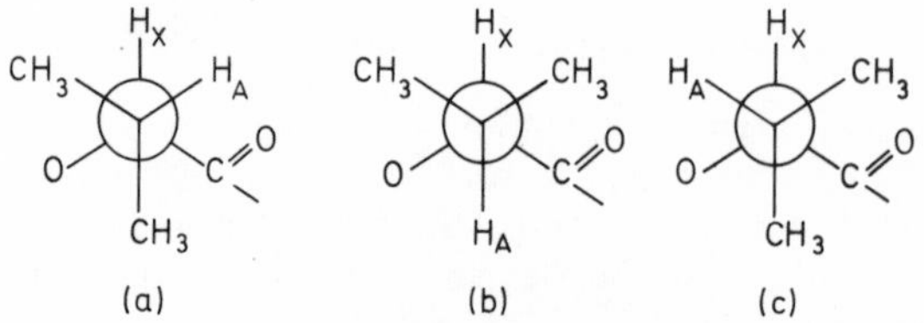

(a)   (b)   (c)

Figure 6. Rotational states of the C^α—C^β bond in the sidechain of the HyIv residue of the diketomorpholines (II)–(IV).

that one of the gauche rotamers a or c is predominant, whereas for compound (II) ($^3J_{\alpha\beta} = 6.2$ Hz) apparently all three or two (a and b or b and c) rotational states are approximately equally populated. The 'shift reagent' Eu(DPM)$_3$ induces a substantial shift of the $C^\beta H_A$ signals in compounds (III) and (IV): -11.5 and -11.4 p.p.m., respectively, practically coinciding with the $C^\alpha H_X$ shifts: -11.6 and -12.0 p.p.m., respectively. Evidently the large $C^\beta H_A$ shifts indicate predominance of the rotamers a of which the H_A proton is situated more closely to the amide carbonyl than in the case of the c rotamer. The differences in the induced shifts of the CH$_3$ signals [-5.7 and -2.3 p.p.m. for (III) and -6.0 and -2.6 p.p.m. for (IV)] also favour the preference for the a rotamer. If the c rotamer were predominant, these signals would have moved downfield for about the same distance.

With diketomorpholine (II), as could have been expected for a levelling of the distribution of the C^α — C^β rotamers, there is a substantially lower $C^\beta H_A$ shift (-8.4 p.p.m.) and an equalizing of the CH$_3$ shifts (-4.6 and -3.5 p.p.m.).

30

It is to be stressed that an addition of Eu(DPM)$_3$ up to a molar ratio of 2:5 does not affect the relative weights of the spatial forms of the molecule. This is evident from the constant integral intensities of the *cis*- and *trans*-isomer signals of dipeptide (I) and from the constancy of the $^3J_{\alpha\beta}$ coupling constants for diketomorpholines (II) to (IV).

IV. ASSIGNMENT OF NH SIGNALS AND DISCLOSING OF 'HIDDEN' LINES IN THE NMR SPECTRA OF PEPTIDES

Undoubtedly an urgent problem in the NMR spectroscopy of peptides is a reliable assignment of the signals. For the proton NMR difficulties arise mainly in the overlap of the C$^\alpha$H and C$^\beta$H signals and in the assignment of the NH doublets to the proper amino acid residue. The most direct way of assigning the NH signals is determination of the multiplicity of the signal from the adjacent C$^\alpha$H group. In this way one may distinguish between amino acid residues with NH—C$^\alpha$H—C$^\beta$H$\diagdown$, NH—C$^\alpha$H—C$^\beta$H$_2$— and NH—C$^\alpha$H—CH$_3$ fragments. Following this there remains to be determined the chemical shift and multiplicity of the C$^\beta$H signal.

The most direct and convenient approach to this problem is the use of the INDOR[37] and detection of signals on combination frequencies[38]. In both techniques one of the radiofrequency fields is swept over the region of the hidden signal. To illustrate their use we shall discuss the interpretation of the proton spectrum of 'symmetric' Val$^{(6)}$,Ala$^{(9)}$–antamanide.

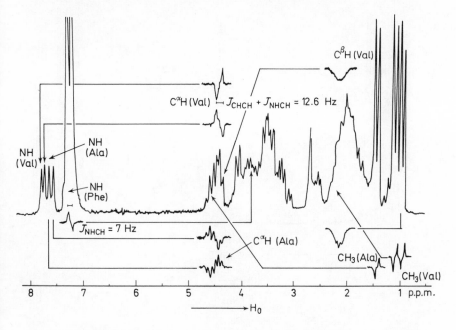

Figure 7. Application of INDOR for signal assignments in the ^{1}H NMR spectrum of 'symmetric' Val$^{(6)}$,Ala$^{(9)}$-antamanide (100 MHz, CDCl$_3$).

Recently INDOR spectroscopy has been used to obtain the ^{13}C spectra of amino acids[29] and to illustrate the coupling of the sidechain CH—CH protons of amino acids and their derivatives[39], and to reduce the sidechain, C^{α} and amino-aromatic proton regions in the peptidic antibiotics spectra[40].

Here we apply the INDOR technique to the assignment of the NH signal in the peptide ^{1}H NMR spectrum (*Figure 7*). On monitoring the components of the low field NH doublet (7.76 p.p.m.), the INDOR response in the $C^{\alpha}H$ region has the appearance of a doublet which is specific for a $C^{\alpha}H$—$C^{\beta}H$ fragment with one β proton. For the given compound this fragment corresponds to the L-Val residues. The separation between the doublet components is the sum of the $^{3}J_{NHCH}$ and $^{3}J_{\alpha\beta}$ coupling constants. On determining the former from the corresponding NH signal one can obtain the latter which cannot be determined from the ordinary spectrum. Further consecutive use of INDOR permits determination of the $C^{\beta}H$ signal and after this of the CH_{3} signals of the L-Val residues.

For the other NH doublet at 7.61 p.p.m. the $C^{\alpha}H$ INDOR response reveals a sextet, corresponding to the $C^{\alpha}H$—CH_{3} fragment, i.e. to the alanyl residue.

The third NH signal stemming from the remaining phenylalanyl residues of the 'symmetric' antamanide is obscured by signals of the aromatic and solvent protons. In order to obtain its INDOR response one must test several peaks in the 3.1 to 4.2 p.p.m. region. Only with two of them at *ca.*

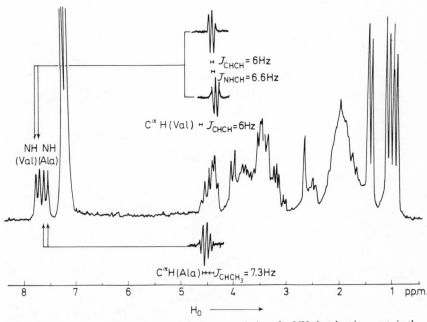

Figure 8. Application of the combination frequency technique for NH signal assignments in the ^{1}H NMR spectrum of Val$^{(6)}$,Ala$^{(9)}$-antamanide (100 MHz, CDCl$_{3}$).

32

3.8 p.p.m. was it possible to obtain the NH INDOR response. From the splitting of this doublet response one could determine the phenylalanyl $^3J_{NHCH}$ coupling constant.

From the viewpoint of the instruments used, the combination frequency signal technique is more complicated than the INDOR technique but it gives a clearer response with higher resolution and sensitivity. *Figure 8* shows the application of this technique for the assignment of the NH signal. When both components of the low field NH doublet are irradiated, the combination frequency $C^{\alpha}H$ response appears as a doublet[38]† which corresponds to the valyl $C^{\alpha}H-C^{\beta}H\diagdown$ fragment. The separation between the response components is directly equal to the $^3J_{\alpha\beta}$ coupling constant of the protons in this fragment. The distances between the responses which are detected on the two alternative combination frequencies ($v_{sweep} - v_1 + v_2$ and $v_{sweep} + y_1 - v_2$, where v_{sweep} is the sweep frequency, and v_1 and v_2 are the irradiation frequencies) is equal to the $^3J_{NHCH}$ coupling constant. The combination frequency response for the other NH signal is a quartet, which is specific for the alanyl $C^{\alpha}H-CH_3$ fragment.

It goes without saying that the INDOR and combination frequency signals techniques do not exclude the simpler spin decoupling. But without doubt they have their advantages in the accuracy and elegance of the experiment.

V. CONCLUSION

The approaches presented above have far from exhausted the capacity of NMR in studies of peptide conformations. The problems next in turn are a detailed analysis of the ^{13}C chemical shifts with respect to conformation, use of relaxation parameters, application of the spin labels and paramagnetic probes, elucidation of the angular dependence of the vicinal $^{15}N-H$ coupling constant for evaluation of the ψ angle and other techniques which are only looming in the background.

REFERENCES

[1] V. F. Bystrov, S. L. Portnova, V. I. Tsetlin, V. T. Ivanov and Yu. A. Ovchinnikov, *Tetrahedron*, 25, 493 (1969).

[2] V. F. Bystrov, S. L. Portnova, T. A. Balashova, V. I. Tsetlin, V. T. Ivanov, P. V. Kostetzky and Yu. A. Ovchinnikov, *Tetrahedron Letters*, 5225 (1969); *Zh. Obshch. Khim.* 41, 407 (1971) (in Russian).

[3] M. Karplus, *J. Chem. Phys.* 30, 11 (1959); *J. Amer. Chem. Soc.* 85, 2870 (1963).

[4] R. J. Abraham and K. G. R. Pachler, *Molec. Phys.* 7, 165 (1963–1964).

[5] L. Pauling, *The Nature of the Chemical Bond*, 3rd ed. Cornell University Press: Princeton (1960).

[6] J. G. Powles and J. H. Strange, *Disc. Faraday Soc.* 34, 30 (1962).

[7] A. J. R. Bourn and E. W. Randall, *Molec. Phys.* 8, 567 (1964).

[8] L. A. La Planche, *Ph. D. Thesis*, University of Michigan (1963).

[9] V. T. Ivanov, I. A. Laine, N. D. Abdullaev, V. Z. Pletnev, G. M. Lipkind, S. F. Arkhipova, L. B. Senyavina, E. N. Meshcheryakova, E. M. Popov, V. F. Bystrov and Yu. A. Ovchinnikov, *Khim. Prir. Soed.* 221 (1971) (in Russian).

---

†The number of the response components is determined as the least of the 'up' and 'down' peaks[38].

[10] Yu. A. Ovchinnikov, V. T. Ivanov, V. F. Bystrov, A. I. Miroshnikov, E. N. Shepel, N. D. Abdullaev, E. S. Efremov and L. B. Senyavina, *Biochem. Biophys. Res. Commun.* **39**, 217 (1970).

[11] V. T. Ivanov, I. A. Laine, N. D. Abdullaev, L. B. Senyavina, E. M. Popov, Yu. A. Ovchinnikov and M. M. Shemyakin, *Biochem. Biophys. Res. Commun.* **34**, 803 (1969).

[12] M. Llines, M. P. Klein and J. B. Neilands, *J. Molec. Biol.* **52**, 399 (1970).

[13] A. Zalkin, J. D. Forrester and D. H. Templeton, *J. Amer. Chem. Soc.* **89**, 1810 (1966).

[14] K. D. Kopple, *Biopolymers*, **10**, 1139 (1971).

[15] V. T. Ivanov, A. I. Miroshnikov, N. D. Abdullaev, L. B. Senyavina, S. F. Arkhipova, N. N. Uvarova, K. Kh. Khalilulina, V. F. Bystrov and Yu. A. Ovchinnikov, *Biochem. Biophys. Res. Commun.* **42**, 654 (1971).

[16] V. T. Ivanov, A. I. Miroshnikov, S. A. Koz'min, E. N. Meshcheryakova, L. B. Senyavina, N. N. Uvarova, K. Kh. Khalilulina, V. A. Zabrodin, V. F. Bystrov and Yu. A. Ovchinnikov, *Khim. Prir. Soed.* in press (in Russian).

[17] J. W. Emsley, J. Feeney and H. L. Sutcliffe, *High Resolution NMR Spectroscopy*, Vol. I, Chapters 8.13.2 and 8.15.1. Pergamon: Oxford (1965).

[18] J. T. Edsall, P. J. Flory, J. C. Kendrew, A. M. Liquori, G. Nemethy, G. N. Ramachandran and H. A. Scheraga, *J. Molec. Biol.* **15**, 399 (1966) and **20**, 589 (1966); *J. Biol. Chem.* **241**, 1004 and 4167 (1966); *Biopolymers*, **4**, 121 and 1149 (1966).

[19] IUPAC–IUB Commission on Biochemical Nomenclature, *J. Molec. Biol.* **52**, 1 (1970); *J. Biol. Chem.* **245**, 6489 (1970); *Biochemistry*, **9**, 3471 (1970); *Europ. J. Biochem.* **17**, 193 (1970).

[20] S. L. Portnova, T. A. Balashova, V. F. Bystrov, V. V. Shilin, Ya. Bernat, V. T. Ivanov and Yu. A. Ovchinnikov, *Khim. Prir. Soed.* 323 (1971) (in Russian).

[21] M. Barfield and M. Karplus, *J. Amer. Chem. Soc.* **91**, 1 (1969).

[22] C. M. Thong, D. Canet, P. Grenger, M. Marraund and J. Neel, *C.R. Acad. Sci., Paris, Sér. C*, **269**, 580 (1969).

[23] R. Schwyzer, private communication cited by R. J. Weinkam and E. C. Jorgensen, *J. Amer. Chem. Soc.* **93**, 7038 (1971).

[24] G. N. Ramachandran, R. Chandrasekaran and K. D. Kopple, *Biopolymers*, **10**, 2113 (1971).

[25] V. Solkan and V. F. Bystrov, *Tetrahedron Letters*, in press (1973).

[26] V. F. Bystrov, V. T. Ivanov, S. A. Koz'min, I. I. Mikhaleva, K. Kh. Khalilulina and Yu. A. Ovchinnikov, *FEBS Letters*, **21**, 34 (1972).

[27] M. Ohnishi, M.-C. Fedarko, J. D. Baldeschwieler and M. F. Johnson, *Biochem. Biophys. Res. Commun.* **46**, 312 (1972).

[28] A. D. Buckingham, *Canad. J. Chem.* **38**, 300 (1960).

[29] W. J. Horsley and H. Sternlicht, *J. Amer. Chem. Soc.* **90**, 3738 (1968).

[30] K. G. Denbigh, *Trans. Faraday Soc.* **36**, 936 (1940).

[31] Yu. A. Ovchinnikov, V. T. Ivanov, A. V. Evstratov, V. F. Bystrov, N. D. Abdullaev, E. M. Popov, G. M. Lipkind, S. F. Arkhipova, E. S. Efremov and M. M. Shemyakin, *Biochem. Biophys. Res. Commun.* **37**, 665 (1969).

[32] C. C. Hinckley, *J. Amer. Chem. Soc.* **91**, 5160 (1969).

[33] L. R. Isbrandt and M. T. Rogers, *Chem. Commun.* 1378 (1971).

[34] S. L. Portnova, V. F. Bystrov, T. A. Balashova, V. T. Ivanov and Yu. A. Ovchinnikov, *Izvest. Akad. Nauk SSSR, Ser. Khim.* 825 (1970) (in Russian).

[35] J. K. M. Sanders and D. H. Williams, *J. Amer. Chem. Soc.* **93**, 641 (1971).

[36] J. Konnert and I. L. Karle, *J. Amer. Chem. Soc.* **91**, 4888 (1969).

[37] E. B. Baker, *J. Chem. Phys.* **37**, 911 (1962).

[38] V. F. Bystrov, *J. Molec. Spectrosc.* **28**, 81 (1968); V. A. Afanas'ev and V. F. Bystrov, *J. Magn. Res.* **3**, 357 (1970).

[39] W. A. Gibbons, H. Alms, J. Song and H. R. Wyssbrod, *Proc. Nat. Acad. Sci., Wash.* **69**, 1261 (1972).

[40] W. A. Gibbons, H. Alms, R. S. Bockman and H. R. Wyssbrod, *Biochemistry*, **11**, 1721 (1972).

VIBRATIONAL SPECTRA OF CONFORMATIONALLY IMPURE POLYMERS

G. Zerbi

Istituto di Chimica delle Macromolecole del CNR, Via Alfonso Corti 12, 20133 Milano, Italy

ABSTRACT

The present status of understanding of the vibrational spectra (infra-red, Raman and neutron scattering) of organic polymers considered as structurally disordered systems is reviewed and discussed. The limitations of the results obtained from the study of polymers as translationally symmetrical systems are pointed out and one of the possible theoretical techniques for the analysis of disordered polymers is presented. It is pointed out that numerical methods are suitable for handling these rather complicated organic molecules and do not require the assumption of structural simplifications which may obscure the real physics of the phenomenon and the corresponding interpretation of the experimental results which are already available. The results obtained so far on polyethylene, polytetrafluoroethylene and hydrogen-bonded systems are discussed.

---

1. INTRODUCTION

In the past decade the main interest of vibrational spectroscopy of polymeric materials has been focused on the understanding of the finger-print region of the spectrum in terms of their molecular structure[1-4]. The largest source of experimental data has been the infra-red spectrum because of the high technological development of the commercial instruments and because of the relative ease in obtaining an infra-red spectrum. The recent availability of laser–Raman spectrometers has provided a sudden burst of experimental data on the Raman spectrum of polymers[5]. While all these data refer to the optical properties, the study of the inelastic coherent or incoherent scattering of neutrons (ICNS, IINS) from these materials has provided additional information on their lattice dynamics and bulk properties[4, 6-8]. The interpretation of these experimental data has been mainly carried out on the basis of an idealized molecular and crystalline model for which it has been assumed that: (i) the polymerization reaction has occurred in an ideal way, (ii) the polymer chain has taken up the most stable conformation as resulting from a balance of intra and inter-molecular forces, (iii) the consequent crystal lattice is perfectly organized. In addition to all these assumptions it has also been considered that the chain has infinite length.

The analysis of the experimental data has been carried out by a large number of authors who have followed three main approaches:

(a) Chemical analysis from group–frequency correlations. This study has proved to be very useful to the chemist in fundamental and practical applications in order chemically to characterize a polymer sample[1, 9-11].

(b) Vibrational analysis using group theoretical methods. This approach

35

has provided valuable information on the structure (stereospecificity and conformation) of the molecular chain[2, 12, 13]. These types of studies have been complementary to x-ray diffraction studies even if sometimes some still unsettled disagreement has been found[14].

(c) Theoretical treatment of the lattice dynamics.

No other class of organic crystals has been the subject of so thorough a treatment as have organic polymers. One of the peculiar reasons is that intra-chain forces of covalent type are at least one order of magnitude larger than inter-molecular (inter-chain) forces of Van der Waals type. In a first approximation the dynamics of one single polymer chain can then be considered neglecting the influence of the neighbouring chains[4]. Only in a second order approximation can inter-chain forces be taken into account from the few observed Davydov's splittings[15-17] or from the (so far unobserved for polymers[18]) static field splittings. Polymer chains can then be considered as 'one dimensional crystals'. This definition implies that we consider a translational symmetry only along the axis of the polymer chain even allowing all the degrees of freedom on three dimensions.

The lattice dynamical treatment is then restricted to the study of phonon waves propagating along the chain axis. Since at the beginning only the optical spectrum was available calculations were restricted to $k = 0$ modes[19-22]. The more recent availability of phonon dispersion curves from inelastic coherent neutron scattering experiments[18, 23-24] and of vibrational density of states[25] from inelastic incoherent neutron scattering experiments has pushed various workers to predict or to verify the dispersion of phonons by a complete lattice dynamical calculation throughout the whole one-dimensional Brillouin zone[4, 26-31]. Theories were proposed and translated into suitable computer programmes which allowed workers to carry out several calculations on the simplest and basically most important polymers. Calculations were also extended to three-dimensional lattices in order to account for Davydov's splitting, for vibrational density of states and dispersion curves for phonons propagating along and across the chain axis[8, 18, 32].

The main contribution given by the studies described in section c is to be able to give a comprehensive interpretation of a large set of experimental data and to derive dynamical properties for the understanding of thermo-dynamical or mechanical properties of these materials. No relevant structural information was derived from these theoretical treatments. However, the availability of theoretical methods has opened new fields in polymer dynamics which we are going to discuss in this paper.

2. PERFECT POLYMERS

The basic starting point of the theoretical treatment is the existence of a translational periodicity or translational symmetry of the lattice which allows us to write the dynamical matrix of the lattice in terms of the properties of the atoms of the unit cell only, thus reducing the complexity of the treatment.

Since we are going to adopt in our discussion mainly the Wilson's type internal displacement coordinates[33], well known to the chemist, let us write the few basic equations to which we shall refer throughout this paper.

We repeat here, for the sake of clarity, the basic theory we have developed and already presented elsewhere[30] for the case of three-dimensional crystals. The reduction to one-dimensional crystals is obvious.

Let us briefly introduce the internal coordinates in the case of the perfect lattice. Let n_1, n_2 and n_3 label each unit cell and let n indicate any given triplet. The nth unit cell contains N atoms whose vibrations can be described by $3N$ Cartesian coordinates collected in the vector x_n or by $3N$ independent internal coordinates collected in a vector r_n. Let t_n be the vector which locates the nth unit cell from a suitably chosen origin. The internal phonon coordinates $R(k)$ can be defined as

$$R(k) = (1/\sqrt{2\pi}) \sum_{-\infty}^{+\infty} {}_n x_n \exp(-ik \cdot t_n) \qquad (1)$$

where k is the wave vector in reciprocal space.

The potential energy of the crystal can be expressed as

$$2V = \int \tilde{R}(k) F_R(k) R(k) \, dk \qquad (2)$$

where the integration is extended over the entire first Brillouin zone. It can be shown that in equation 2 $F_R(k)$ takes the form

$$F_R(k) = F_R^0 + \Sigma_s \tilde{F}_R^s \exp(-ik \cdot t_s) + F_R^s \exp(ik \cdot t_s) \qquad (3)$$

where F_R^0 is the matrix of the force constants relating coordinates within the same cell; F_R^s is the matrix of the force constants relating the coordinates of two cells s units apart.

Even if the dynamical matrix of the perfect crystal can be written in terms of the $R(k)$ using equation 3 and a proper metrical tensor $G_R(k)$, we prefer in this case to express it in terms of the usual Cartesian phonon coordinates

$$X(k) = (1/\sqrt{2\pi}) \Sigma_r \, x_n \exp(-ik \cdot t_n) \qquad (4)$$

We must then look for the transformation matrix $B(k)$ relating the Cartesian phonon coordinates to the internal ones. Using Wilson's technique the internal coordinate vector r_n can be expressed as

$$r_n = B_{-l} x_{n-l} + \ldots + B_1 x_{n-1} + B_0 x_n + B_1 x_{n+1} + \ldots + B_l x_{n+l} \qquad (5)$$

$$= B_0 x_n + \sum_{l}^{m} (B_{-l} x_{n-l} + B_l x_{n+l}) \qquad (6)$$

where x_{n+l} indicates the Cartesian coordinate vector of a unit cell, l units distant from the nth cell. B_l is the corresponding transformation matrix between r_n and x_{n+l}. Substitution of equation 6 into equation 1 and taking account of equation 4 gives

$$R(k) = \Sigma_l^{m} {}_{-m} B_l \exp\{-ik \cdot t(l)\} X(k) \qquad (7)$$

37

from which the desired transformation matrix $B(k)$ can be derived:

$$B(k) = \sum_{-m}^{m} B_l \exp\{-i k \cdot t(l)\} \tag{8}$$

We can now express the potential energy in Cartesian coordinates in terms of the potential energy in internal coordinates as:

$$F_x(k) = \tilde{B}(k)F_R(k)B(k) \tag{9}$$

from which the dynamical matrix is derived

$$D(k) = M^{-\frac{1}{2}}F_x(k)M^{-\frac{1}{2}} \tag{10}$$

M being the diagonal matrix of the masses.

The $3N$ dispersion curves for a given crystal can be obtained numerically from the solution of the eigenvalue equation

$$[D(k) - \omega^2(k)I]L_x(k) = 0 \tag{11}$$

once the force field F_R is numerically known. The eigenvectors $L_x(k)$ can also be obtained from the solution of equation 11. Throughout this paper it will be assumed that the force field in internal coordinates is well known†.

The previous theory has been applied by us or by other authors to several polymeric materials. The reader is referred to the corresponding literature for a detailed discussion of the results and of the problems still left unsolved for each particular polymer molecule.

With the availability of large and fast computers calculations have been extended to more and more complicated cases. While the lattice dynamics of perfect polyethylene and polyoxymethylene can at present be considered satisfactorily known[4, 8], isotactic polypropylene, polyvinyl chloride and several other basic polymers still require further studies. Recently calculations have been extended to models of polypeptides in order to tackle in a more quantitative and detailed way the extremely difficult problem of biopolymers[34–36]. We now consider it useful to make a few general comments:

(1) Because of the large number of atoms per chemical repeating unit the number of phonon dispersion branches becomes extremely large. In practice a continuous set of phonon waves can be transmitted by the polymer throughout the whole frequency range from 0 to approximately $1\,800$ cm^{-1} with very little or no energy gaps.

(2) Several attempts have been made to reduce the difficulty of the problem by reducing the complex polymer chain to a model of the most meaningful point masses[34–35]. The results of these types of studies should be treated with great caution since the reduction to a point mass model in general forcefully neglects dynamically important variables thus altering the validity of the derived conclusions. The discrepancy of the results of the studies of solid hydrogen-bonded polymers like methanol[37–39] and ice[40–44] can be taken as typical examples.

(3) Several phonon branches are calculated to be independent of k for

---

† For a critical discussion on force fields for polymers, see ref. 4.

the following main reasons: (i) strong covalent forces do not allow strong couplings between neighbouring oscillators, (ii) groups of atoms in side-chains are dynamically uncoupled (either through potential or kinetic energy) between each other.

(4) When forces become weaker (this generally occurs for bending and torsional motions) a large dispersion takes place and phonons belonging to the same symmetry species for a given k couple giving rise to non-crossing between phonon branches with the consequent mixing of the various oscillators. An overall description of the modes derived from potential energy distribution[45] becomes then impracticable. It has to be pointed out, however, that the symmetry of polymer chains is high only for very simple systems (e.g. polyethylene, tellurium, selenium, polyoxymethylene, poly-tetrafluoroethylene) when considered at the Γ point. The symmetry of phonons for $k \neq 0$ is much lowered for most of the cases.

3. IMPURE AND DISORDERED POLYMERS

We wish to treat here in more detail the problem of the lattice dynamics of polymeric materials when some sort of defect or disorder is introduced into the chain.

It is now of common knowledge that the idealized model of a polymer chain is only a very rough approximation of the real polymer. Through several independent physicochemical techniques the presence of amorphous or irregular or disordered material in a solid sample of a given polymer has been proved[46-49]. The various types of defects which can possibly occur in a polymer have already been discussed[50]; they can then generally be collected into four main classes: *chemical defects* (isotopic substitution, switching of the direction of insertion of the monomer unit into the chain, chain branching, crosslinking); *conformational defects* (chain folding, kinks, jogs); *stereochemical defects* (different tacticity) and *packing defects* (amorphous phase, lack of three-dimensional order).

The vibrational spectrum (infra-red, Raman and neutron scattering) shows several non-negligible and sometimes very prominent features which should be interpreted as arising from the imperfect part of the chain. The analysis of these extra features can be carried out in two ways, namely on qualitative grounds using empirical spectral correlations[51] or by a quantitative approach based on lattice dynamical calculations. While the former approach has been widely adopted in industrial laboratories for a quick and qualitative characterization of a polymeric material[11], we have recently carried out some studies based on a detailed quantitative approach[52-56].

The dynamical quantities from experiments or calculation on the perfect lattice which we consider of basic importance for a quantitative study of the problem of real polymers are the following:

phonon dispersion curves, $\omega(k)$;
optical phonon frequencies, $\omega(k = 0)$;
optical phonon intensities, $I(\omega)$, $\omega(k = 0)$;
one phonon density of states, $g(\omega)$;

amplitude weighted one-phonon density of states, $g(\omega)^A$;

dipole or polarizability weighted density of states, $g(\omega)^\mu$.

As will be discussed in this paper some of the dynamical quantities just mentioned can be obtained readily from a standard lattice dynamical calculation on *perfect one-dimensional* or three-dimensional polymeric materials using equations 1 to 11. A few require more careful consideration and are now the subject of extensive studies in our laboratory. The $I(\omega)$ determine the appearance of the optical spectrum and are influenced by the symmetry of the system. The higher the symmetry the more strict are the operative optical selection rules and thus substantially simplify the optical spectrum (e.g. polyethylene[28]). $\omega(k)$, $g(\omega)$, $g(\omega)^A$ mainly determine the experimental results from neutron scattering experiments on single crystals, on stretch-oriented or on polycrystalline materials.

The introduction of some sort of defect into the model modifies the dynamics of the system to an extent depending on the concentration of defects and on their coupling with the host lattice or with neighbouring defects. Several efforts have been variously made by many authors for the analytical treatment of the effect of a defect on the dynamics of a defect-containing lattice. The basic concept is to treat the perturbation to the dynamics of the perfect lattice by the defects using the Green's function method[57–59]. The basic limitation of this method is that the lattice should be quite simple, the defect perturbation be small (e.g. isolated mass and force constant defects) and their concentration small such that the defects cannot interact. The results of these analytical treatments have been mainly applied and verified in the case of simple inorganic crystals[60, 61].

One of the peculiar features of polymeric systems is that the concentration of defects is generally high and the defects entail changes of geometry and force constants throughout several atoms along the chain. One thus moves from an *impure* system to a disordered one. It then becomes impossible to apply the Green's function method to real polymeric systems and numerical methods must be developed for this purpose[62, 63].

The main reason why a numerical method becomes useful is that a quantity of direct physical meaning required from such dynamical systems is the density of vibrational states $g(\omega)$. While the knowledge of a single eigenvalue may be of no use or only very rarely required, the knowledge of the whole vibrational spectrum is compulsory for the understanding of the physical quantities related to vibrations. On the same grounds it is not essential to derive the complete set of eigenvectors but one may need only a few which may be essential in the analysis of the experimental data such as infra-red, Raman and neutron scattering spectra.

The numerical method based on the Negative Eigenvalue Theorem (NET) we are going to discuss allows the direct calculation of $g(\omega)$ at any desired accuracy thus reaching, if necessary, that of a single isolated eigenvalue. Approximate or exact eigenvectors can also be obtained by the so-called 'inverse iteration method'. The NET was introduced first by Dean et al.[63] who treated the vibrations of several disordered systems in this way.

Let us describe briefly the main lines of the procedure. We wish to compute the number $n(\omega_2 - \omega_1)$ of eigenvalues of the $3Np \times 3Np$ dynamical matrix D which lie in the interval (ω_1, ω_2) where ω_1 and ω_2 are positive real numbers such that $\omega_2 > \omega_1$. The number $n(\omega_2 - \omega_1)$ is given by

$$n(\omega_2 - \omega_1) = \eta(D - \omega_2 I) - \eta(D - \omega_1 I) \qquad (12)$$

where I is the $3Np \times 3Np$ unit matrix and $\eta(D - \omega_i I)$ is the number of negative eigenvalues of the matrix

$$D_i = D - \omega_i I \qquad (13)$$

The computation of the negative eigenvalues of D_i is performed by a particular partitioning of D_i, which applies to any symmetrical matrix. This is particularly convenient since D has a codiagonal form.

The NET states that, given a symmetrical matrix M of dimensions $r \times r$, partitioned as follows

$$M = \begin{bmatrix} A_1 & B_2 & & \bigcirc \\ \tilde{B}_2 & A_2 & B_3 & \\ & & \ddots & \\ \bigcirc & & \tilde{B}_k & A_k \end{bmatrix} \qquad (14)$$

where A_i has dimensions $r_i \times r_i$
$\quad\quad B_i$ has dimensions $r_{i-1} \times r_i$

and $\quad \sum\limits_{i+1}^{k} r_i = r$

then the number $\eta(M - xI)$ of negative eigenvalues of the matrix $M - xI$ is given by

$$\eta(M - xI) = \sum_{i=1}^{k} \eta(U_i) \qquad (15)$$

where

$$\begin{aligned} U_i &= A_i - xI_i - \tilde{B}_i U_{i-1}^{-1} B_i \\ U_1 &= A_1 - xI_1 \end{aligned} \qquad (16)$$

The particular partitioning we perform on D_i is the following. Let $D_i^{(1)}$ denote the matrix D_i partitioned as

$$D_i^{(1)} = \begin{matrix} X_1 & Y_1 \\ \tilde{Y}_1 & Z_1 \end{matrix} \qquad (17)$$

where $X_1 \quad$ is a 1×1 matrix
$\quad\quad Y_1 \quad$ is a $1 \times (3Np - 1)$ matrix
$\quad\quad Z_1 \quad$ is a $(3Np - 1) \times (3Np - 1)$ matrix

Then from equation 15, one has

$$\eta(D_1) \equiv \eta(D_i^{(1)}) = \eta(X_1) + \eta(D_i^{(2)}) \qquad (18)$$

$$D_i^{(2)} = Z_1 - \tilde{Y}_1 X_1^{-1} Y_1 \qquad (19)$$

One can continue the process as indicated in equation 19 for $3Np - 1$ times until one reaches the result that

$$\eta(\boldsymbol{D}_i) = \sum_{j=1}^{3Np} \eta(\boldsymbol{X}_j) \tag{20}$$

This particular partitioning avoids inversion of matrices because each $\boldsymbol{X}_i$ is a 1×1 matrix. The time for computing each $\boldsymbol{D}_i^{(k)}$ is lowered by the fact that the row matrix $\boldsymbol{Y}_{k-1}$ entering in its definition has only $C - 1$ non-zero elements where C is the number of codiagonals of $\boldsymbol{D}$; thus the elements of $\boldsymbol{Z}_i$ which are modified by equation 19 are only those lying in its first $(C - 1)$ rows and columns†.

The intervals (ω_1, ω_2) can be restricted to any desired accuracy. For comparison of the calculated $g(\omega)$ with the spectra it is enough to use

$$\omega = \omega_2 - \omega_1 = 5 \, \text{cm}^{-1}$$

but for the computation of eigenvectors a more precise knowledge of the approximate eigenvalues is required. For this purpose the interval $d\omega = \omega_2 - \omega_1$ in the region of the spectrum where we wish to know the eigenvectors is restricted more and more until $d\omega$ contains only one eigenvalue. Let us call $\bar{\omega}_i$ the value $\frac{1}{2}(\omega_2 - \omega_1)$ and ω_i the (unknown) exact eigenvalue of $\boldsymbol{D}$ occurring in $d\omega$. The procedure applied to compute ω_i ensures that

$$\omega_i - \bar{\omega}_i < \omega_k - \bar{\omega}_i \tag{21}$$

where ω_k is any other exact eigenvalue of $\boldsymbol{D}$. For computing the eigenvector $\boldsymbol{L}_i$ associated with ω_i we now proceed with the 'Inverse Iteration Method'[64]. We choose an arbitrary trial $3Np$ vector $\boldsymbol{u}_0$ and construct

$$\boldsymbol{v}_1 = (\boldsymbol{D} - \omega_i \boldsymbol{I})^{-1} \boldsymbol{u}_0 \tag{22}$$

$$\boldsymbol{u}_1 = \boldsymbol{v}_1 / \text{max.}(\boldsymbol{v}_1)$$

where max.$(\boldsymbol{v}_1)$ denotes the element of largest modulus in $\boldsymbol{v}_1$ and then repeat for a certain number of times s the two steps 22 of iteration. The rapidity of convergence of $\boldsymbol{u}_s$ to $\boldsymbol{L}_i$ (the true eigenvector) depends on the condition 21. Let us express the arbitrary vector $\boldsymbol{u}_0$ by means of the complete set of eigenvectors $\boldsymbol{L}_k$ of $\boldsymbol{D}$

$$\boldsymbol{u}_0 = \sum_{k=1}^{3Np} \alpha_k \boldsymbol{L}_k \tag{23}$$

Then any step in equations 22 can be re-written, apart from a normalization factor,

$$\boldsymbol{u}_s = \sum_k \alpha_k (\omega_k - \bar{\omega}_i)^{-s} \boldsymbol{L}_k$$

Equation 21 ensures that the contribution of $\boldsymbol{L}_k (k \neq i)$ to $\boldsymbol{u}_s$ decreases with increasing s‡.

---

† The time required to compute a step of the histogram for $g(\omega)$, as given in equation 12, for a matrix $\boldsymbol{D}$ with dimensions $1\,800 \times 1\,800$ and 30 non-zero codiagonals is about 30 seconds on a UNIVAC 1106.

‡ By comparison with the Jacobi method on 120×120 matrices three iterations are enough to ensure convergence up to the sixth decimal place, when equation 21 is satisfied.

The use on computers of equation 20 is difficult because large matrices must be inverted. This difficulty can be avoided[12] by using the following decomposition of $D - \bar{\omega}_i I$

$$D - \bar{\omega}_i I = \tilde{U} \varDelta U \tag{24}$$

where U is an upper triangular matrix and the elements of the diagonal matrix $\varDelta$ are chosen in such a way that the diagonal elements of U are unity. By use of equation 24 each step in the iteration can be split into three equations

$$\begin{aligned} \tilde{U} Y_s &= \mathbf{u}_s \\ \mathbf{v}_{s+1} = (D - \omega_i I)^{-1} \mathbf{u}_s \qquad \varDelta z_s &= \mathbf{v}_s \\ \tilde{U} \mathbf{v}_{s+1} &= \mathbf{z}_s \end{aligned} \tag{25}$$

each of which is very easy to solve because of the particular form of the matrices involved†.

For a better understanding of the complicated pattern of motion of defect-containing lattices we usually compute also the eigenvectors on the basis of internal or group coordinates[13].

General remarks

$g(\omega)$ calculated with the above methods is the basic dynamical quantity to be used for the interpretation of the optical (infra-red and Raman) and neutron scattering experimental data. A direct comparison of $g(\omega)$ can be carried out only with the density of states derived from IINS experiments since for unoriented samples there is no need to take into account polarization vectors (i.e. atomic displacements) and the Debye Waller factor can be considered a second order correction to the calculated spectrum.

The comparison with the optical spectrum requires the knowledge of the transition moment for each normal mode of such a complicated atomic system. The general concepts derived from the treatment of impure lattices[60, 65] can be applied also to polymers and can also be extended when the concentration of the defects becomes larger or the type of defect is such that the vibrational perturbation becomes sizeable. We neglect in this discussion the problem of *out-of-band* or *gap-modes* which, with certainty, occur in simple systems containing mass defects and concentrate our attention on defects of geometry. A detailed discussion of the problem of a large concentration of mass defects has been presented in detail elsewhere for the case of isotopic mixed crystals[66, 67].

For the case of organic polymers geometrical defects are more important and have been the subject of several published works.

As already pointed out in section 2 of this paper for organic polymers, even in a perfect state, the whole energy range from 0 cm^{-1} to the near infra-red region is practically covered in a continuous way by dispersion curves. Any extra mode arising from a defect introduced in the perfect host chain will be coupled to the phonons of the host lattice by an extent depending

---

† The time required to compute an eigenvector of a matrix 900×900 with 30 codiagonals on UNIVAC 1106 is about seven minutes.

on the dynamical conditions (geometry and potential). We have then to consider only *resonance modes* and their dependence on concentration, on geometry and on force field.

For comparison with optical spectra we may use the following criteria:

For a low concentration of defects: (Impure System)
(a) $k = 0$ modes from the host lattice allowed by symmetry will be primarily observed.
(b) Because of the breakdown of translational periodicity $k \neq 0$ phonons from the host lattice will gain some spectral activity. The density of state of the host lattice will be somehow mapped in the optical spectrum.
(c) Resonance modes of a particular defect may give rise to additional peaks which become characteristic of the defect and may be used for the diagnosis of the existence and concentration of such a defect in an unknown polymer sample.

Effects a to c can be predicted with fair certainty by the numerical calculations previously presented only if we are concerned with their frequency or shapes of modes. The knowledge of the optical transition moments which modulate the $g(\omega)$ is still an unsolved problem which requires careful and extensive studies[66, 67]. Most of the works so far carried out have explicitly assumed a dipole or polarizability unweighted density of states[52-56]. Comparison between optical intensity and height of the peaks in $g(\omega)$ could not in principle be carried out. In a first approximation, however, it has been assumed that, since the geometry of an actual polymer sample is far from being ideal and a sizeable amount of disorder does exist, any restriction by symmetry is removed and all modes gain some activity. Hence coincidence between peaks in the spectrum and in the calculated $g(\omega)$ can in a rough analysis be taken as positive coincidences[50]. Further works on model compounds and on simpler systems are required for substantiating the conclusions derived from the comparison of the optical spectrum with dipole or polarizability unweighted density of states.

When the concentration of geometrical defects increases the analysis must consider the problem of *disordered lattices*. The concept of resonance modes and phonons of the host lattice is lost and each normal mode becomes the result of a complex coupling between oscillators[67]. For such disordered systems the numerical method is the only one which can provide physically meaningful information from the calculated $g(\omega)$ and the corresponding calculated amplitudes. If some sort of charge model is assumed spectral intensity can also be predicted[66, 67]. For disordered polymers the following general criteria should be considered:

(a) While the density of vibrational states of a disordered system cannot be compared in principle with that of the corresponding perfect lattice, some indication of residual order in the disordered network can still be derived by the experimental finding of $g(\omega)$ peaks corresponding to the $k = 0$ modes of the perfect lattice. The existence of quasi-phonon waves in quasi-ordered regions of a disordered lattice requires further experimental and theoretical studies. The main question which arises is the following: how large should an ordered cluster be to give rise to quasi-phonon waves? This problem is of particular importance for the understanding of the vibrational spectra of

44

co-polymers or block-polymers. It is well known that empirical, correlative or semiquantitative analysis of the vibrational spectrum has been the basic tool for the characterization of polymers and co-polymer structure and composition in applied and industrial research. These empirical studies require a more quantitative theoretical support.

(b) The study of lattice dynamics of disordered polymers acquires a particular importance in the field of vibrational spectroscopy of biopolymers. In fact the degree of ordering, the 'blockness' and the regularity of these substances is still a basic structural problem which needs to be clarified. Furthermore since most of these substances cannot be crystallized a large amount of disordered or irregular material is likely to exist.

(c) A relevant result from the numerical calculations based on NET is that the calculated histogram of the density of states often shows a fine structure or a large population of peaks. The identification of the origin of each peak through the calculation of the corresponding eigenvectors allows us to assign some of these peaks to islands or blocks or segments of particular dimensions[67]. Furthermore some of the peaks may turn out to be characteristic of a particular defect or of a cluster. If the experiments fit the calculation it is thus possible to probe with the vibrational spectrum the inside of the disordered material with regard to its structure, conformation and configuration. A clear example of 'island analysis' and its usefulness has been presented for a realistic case of isotopic mixed crystals[67]. The same kind of analysis has been carried out on models of three-dimensional lattices[62]. It has to be pointed out that no analytical treatment is able to provide such a detailed set of theoretical data to be compared with experiments.

(d) The possibility of such an 'island analysis' invites a yet unexplored theoretical treatment of co-polymers and block polymers.

4. APPLICATIONS

Polyethylene (PE)

(1) The study of the infra-red spectra of copolymers of ethylene and deuteroderivatives of ethylene on the basis of the dynamics of mass disordered systems has unequivocally proved that the polymerization of ethylene with a Ziegler type catalyst occurs with the *cis* opening of the double bond[52]. Miyazawa had previously reached the same conclusion on somewhat weaker spectroscopic evidence[68].

(2) The microstructure of solid PE, i.e. the conformation the chain takes up in the solid state, had been studied[53] with the methods previously discussed in this paper. While a very large amount of experimental and theoretical studies on the dynamics of a *perfect* model of PE has been carried out[4, 8] several non-negligible features in these spectra were left unexplained and qualitatively ascribed to the so-called 'amorphous' part of the polymer[2] which is known to exist even in the case of single crystals.

Since the solid state of a polymeric material is generally described as that of crystallites of various sizes non-homogeneously distributed and embedded in an amorphous substance[46] we have generated long planar zig-zag sequences of CH_2 units intermingled with regions of coiled or geometrically disordered chain segments. Such a structure has been translated into a dynamical matrix built by appropriate mathematical devices[53]. $g(\omega)$ was then derived using NET. The markovian structure of the chain has been accounted for by a proper choice of the 'allowed' geometrical defects, i.e. of the 'allowed' simplest combinations of the most common internal rotational angles. Geometrical defects of the types G, GTG, GTG', GTTG, GGTGG etc. were introduced into the polymer chain while all the other CH_2 units were held in T conformation. The following results were obtained:

(i) The existence of defects results in the activation in the Infra-red or Raman of several singularities in the $g(\omega)$ of the host lattice which should be inactive for the perfect crystal. There is a general activation of all band modes.

Recently Gall et al.[69] have suggested an alternative explanation of the Raman band at 1461 cm^{-1} which we have instead ascribed to the activation of a $k = 0$ phonon only infra-red active for the perfect case. These authors suggest that the Raman line arises from two-phonon processes of the perfect lattice whose intensity is enhanced by some sort of Fermi resonance supposedly with the $k = 0$ fundamental at 1440 cm^{-1}. The interpretation suggested by Gall et al. does not account for the fact that the Raman line at 1461 cm^{-1} depends on the content of amorphous material[70].

(ii) Peaks of $g(\omega)$ characteristic of specific defects of the disordered section of PE have been predicted and experimentally verified. Of particular interest are the two resonance modes at 1350 and 1365 cm^{-1} assigned to GG + GTG and GTTG defects respectively. Their relative intensity changes with temperature, even before melting. It is thus possible to follow the thermal history of the sample (i.e. the structural evolution with temperature) of a solid sample of PE even before melting. It is then possible to have some indication that the melting process consists in an increase of concentrations of kinks and folds within the chain of the type GTG and GG or GTG' at the expense of all-*trans* of GTTG segments. This interpretation is also supported by additional theoretical and experimental evidence from infra-red and neutron scattering data in the lower energy region. As previously discussed in the spectrum of liquid PE it is possible to find indication of $k = 0$ phonons of the perfect lattice thus indicating that the chains are not completely coiled in the liquid state but keep a certain degree of quasi order which generates quasi phonon waves.

(iii) It has been possible to extend the analysis to the case of single crystals grown from solution. The interpretation of the infra-red spectrum based on NET cannot disclaim the existence of some tight fold re-entry at the surface of the single crystal[72] (GGTGG fold) but supports the existence of additional disordered material confirming the so-called disordered or 'composite-fold' model[47].

On the far infra-red spectrum of liquid paraffins and nujol

The far infra-red spectrum of liquid octane, decane and hexadecane has been recently reported by Hall et al.[73]. The same authors report on the temperature dependence of the far infra-red spectrum of nujol from the liquid to the solid state. They observe a main band between 250 and 300 cm^{-1} flanked by a weaker very broad band with maximum at ~ 150 cm^{-1}. Based on the fact that the spectrum of nujol decreases in intensity with decreasing temperature these authors conclude that the observed far infra-red spectrum of liquid paraffins is mainly due to two-phonon processes. Support for their interpretation comes from calculations based on planar zig-zag chains of CH_2 of various lengths with CH_2 taken as a point mass. Summation processes are carried out and the intensity is calculated on the basis of two-phonon interaction theory by Lax and Burstein[74].

From the results of our studies on geometrically disordered paraffins we wish to suggest an alternative explanation of the spectra of liquid paraffins and nujol and its temperature dependence. It is unlikely that paraffins in the liquid state keep the *trans* planar conformation assumed by the previous authors, but quite likely they are randomly coiled thus approaching the model of the disordered paraffins studied by us and discussed in the previous section. The density of states of the perfect long chain paraffin shows two cut-off frequencies at ~ 500 and ~ 200 cm^{-1} corresponding to the highest flat points of in-plane deformation and out-of-plane or torsional branches v_5 and v_9. If a spectral activation of one-phonon transitions is accepted the very weak peak at ~ 150 cm^{-1} is accounted for. The far infra-red spectrum of solid as well as liquid polyethylene shows two broad bands which we assign in a similar way to the two peaks of one-phonon density of states. The calculation of the density of states of all *gauche* polyethylene[53] in addition to two peaks at 100 and 600 cm^{-1} shows a broad strong peak centred at ~ 320 cm^{-1} which is characteristic of G segments. We suggest that the absorption between 250 and 300 cm^{-1} observed by Hall *et al.* for liquid paraffins and nujol arises from *gauche* segments which are very likely to exist in the liquid phase. Unquestionable evidence of the introduction of other conformations in going from the solid to the liquid paraffin comes from the work by Schonhorn and Luongo[75] who observe that at the melting temperature a doublet at 1 350 and 1 365 cm^{-1} clearly appears in the infra-red. Just at the same frequencies we calculate resonance modes assigned to GG + GTG and GTTG segments. We then conclude that the infra-red peak of liquid paraffins and nujol between 250 and 300 cm^{-1} may arise from funda-mental transitions coming from coiled sections of the paraffin molecules. The temperature dependence of the 250 cm^{-1} band of nujol can simply be accounted for as a continuous structural evolution of the paraffin mixture which tends to decrease the concentration of coiled chains by lowering the temperature. It should be pointed out that while solid polyethylene does not show any absorption in the 300 cm^{-1} region, molten polyethylene shows a peak at ~ 315 cm^{-1} which we have already assigned to *gauche* or *quasi-gauche* segments[53]. The precise determination of the frequencies is not pos-sible because (1) the force field is approximate, (2) the real conformation of liquid paraffin or polyethylene may not be precisely G ($\pm 120°$) but may take up slightly distorted values.

Polytetrafluoroethylene (PTFE); Spectrum and phase transitions

From several physicochemical techniques it has been shown that solid PTFE may exist in different phases. The identification of the precise structure of these phases and the mechanism of the phase transition by x-ray n.m.r., dielectric and thermodynamical studies is somewhat confused and sometimes conflicting.

It is experimentally certain that PTFE has two main phase transitions at 19°C and 30°C. Some x-ray studies have established that below 19°C the PTFE chain coils into a translationally regular helix whose identity period of 16.8 Å contains 13 CF_2 groups in six turns[76]. Phase transitions are qualitatively ascribed to the onset of some structural disorder variously described by different authors[77]. Recent conformational studies[78] based on semi-empirical two-body interaction potentials predict that the most stable chain structure corresponds to a 13/6 helix with rotational angle of $\tau = \pm 165.66°$ in good agreement with the x-ray result. The two minima at $+166°$ and $-166°$ are divided by a very low potential barrier at $\tau = 180°$. Other less stable minima are predicted at $\sim \pm 91°$ and $\sim \pm 65°$ corresponding to $\sim 10/3$ and $\sim 4/1$ helices respectively. Because of the very low potential barrier through $\tau = 180°$ Giglio et al.[78] tentatively suggest that above 19°C the chains might consist of a mixture of segments of left and right handed helices joined through bonds in *trans* conformation ($\tau = 180°C$).

We have then recently treated[55,56] the case of the disordered PTFE chain in the hope of contributing to the understanding of the structural features of this important polymer. Indeed the infra-red[2], Raman[79,80] and ICNS spectra[18] show several features which cannot be explained in terms of 13/6 or 15/7 helices. $g(\omega)$ was calculated for perfect and conformationally disordered chains using NET. No dipole weighting has been performed, but consideration of intensity in the infra-red and Raman were taken into qualitative consideration in the comparison of the theoretical data with the experiments. From our studies the following conclusions were derived:

(i) At room temperature there exists a non-negligible amount of *trans*-planar segments as predicted from conformational energy calculations. The phase transition at 19°C corresponds to a sudden increase in concentration of *trans*-planar segments.

(ii) Evidence is found for the existence of another type of helix with geometry close to that of the 10/3 helix. No evidence is found for the existence of a helix with geometry close to 4/1.

(iii) Most of the spectral features which remained unexplained from a vibrational analysis based on perfect 13/6 or 15/7 models were accounted for.

(iv) The evidence of Davydov's splitting in the Raman lines claimed by Koenig et al.[81] and the consequent possibility of the existence of a unit cell containing more than one chain has been shown to be not unequivocal. Most of the temperature dependent doublets can be reasonably ascribed to a conformational equilibrium which changes with temperature.

(v) The recent additional data from coherent neutron scattering experiments[18] which could not be clearly explained in terms of a perfect 13/6 helix can be easily ascribed to segments of 10/3 helix which is likely to exist in most of the samples.

(vi) The disorder probed in PTFE by the vibrational spectrum is related only to that which occurs within the chain. Amorphous sections due to the lack of three-dimensional order only are not detected; contrary to what was previously assumed by many authors, no 'amorphous band' may be located in this way.

Disordered hydrogen bonded systems

The type of study so far discussed in this paper has been extended recently to the case of the simplest hydrogen bonded networks in the solid state namely to the cases of hydrochloric acid $(HCl)^{67}$, ice$^{43, 44}$ and methanol^{39}. It should be pointed out that an unequivocal understanding of the dynamics of these seemingly very simple systems is hindered by the fact that disordered phases may exist.

The combination of the analysis of the spectra based on perfect as well as disordered models has allowed a more detailed understanding of the spectrum of ice Ih$^{43, 44}$. While Shawyer and Dean have already reported on the calculation by NET on a three-dimensional finite crystal of ice^{82} we have carried out the lattice dynamical analysis on a complete perfect model which includes the hydrogen atoms and four molecules per unit cell. The works by previous authors were instead mainly based on highly simplified models$^{40-42}$. On the basis of our theoretical calculations compared with the experiments the evidence claimed by other authors of the existence of long range electrostatic forces^{40} cannot be accepted. In a similar way we cannot support the conclusions of the existence of fluctuating polar ordered domains^{41} or of O—O bonds of different nature perpendicular and oblique to the c-axis of the ice crystal$^{41-42}$.

In the case of solid methanol the phase transition between the α phase stable below and a β phase above the transition temperature of 157.4 K have been studied. The combined results of CNDO, packing energy and lattice dynamical calculations^{39} suggest that the α phase is predominantly ordered and that the second order transition at 157.4 K is mainly caused by the onset of flip-flop motion of the CH_3 groups below and above the plane formed by the oxygen atoms. The structure of the β phase is disordered and accounts for several independent physicochemical measurements which were variously interpreted by several authors^{39}.

ACKNOWLEDGEMENTS

I wish to express my gratitude to Dr M. Gussoni for her continuous help and for helpful discussions throughout the whole work on defects.

REFERENCES

[1] J. Dechant, *Ultrarot-spektroskopische Untersuchungen an Polymeren* Akademie-Verlag: Berlin (1972).

[2] S. Krimm, *Fortschr. Hochpolymer. Forsch.* **2**, 86 (1960).

[3] A. Elliott, *J. Polymer Sci.* **C7**, 37 (1964).

[4] G. Zerbi, *Applied Spectroscopy Reviews*, Vol. II, edited by E. G. Brame. Marcel Dekker: New York (1969).

[5] J. L. Koenig, *Applied Spectroscopy Reviews*, Vol. IV, edited by E. G. Brame. Marcel Dekker: New York (1971).

[6] H. Boutin and A. Yip, *Molecular Spectroscopy with Neutrons*, MIT Press: Cambridge, Mass. (1968).

[7] L. Holliday and J. W. White, *International Symposium on Macromolecules, Leiden (1970)*; Butterworth: London (1971).

[8] T. Kitagawa and T. Miyazawa, *Advances in Polymer Science*, Springer: Berlin (1972).

[9] D. O. Hummel, *Infrared Spectra of Polymers in the medium and long wavelengths region*, Interscience: New York (1966).

[10] C. Tosi and F. Ciampelli, *Adv. Polymer Sci.*, in press.
[11] C. Tosi and G. Zerbi, *Chimica e Industria*, 55, 334 (1973).
[12] G. Zerbi, F. Ciampelli and V. Zamboni, *Chimica e Industria*, 46, 1 (1964);
G. Zerbi, F. Ciampelli and V. Zamboni, *J. Polymer Sci.*, Part C, 7, 141 (1964).
[13] G. Cortili and G. Zerbi, *Spectrochim. Acta*, 23A, 285 (1967).
[14] G. Zerbi and G. Cortili, *Spectrochim. Acta*, 26A, 733 (1970).
[15] H. A. Willis, R. G. Miller, D. M. Adams and H. A. Gebbie, *Spectrochim. Acta*, 19, 1457 (1963).
[16] V. Zamboni and G. Zerbi, *J. Polymer Sci.*, Part C, 7, 153 (1964).
[17] G. Zerbi and G. Masetti, *J. Molec. Spectrosc.* 22, 284 (1967).
[18] L. Piseri, B. M. Powell and G. Dolling, *J. Chem. Phys.* 58, 158 (1973).
[19] J. H. Schachtschneider and R. G. Snyder, *Spectrochim. Acta*, 20, 853 (1964).
[20] T. Miyazawa and Y. Ideguchi, *Bull. Chem. Soc. Japan*, 37, 1065 (1964).
[21] H. Todokoro, A. Kobayashi, Y. Kawaguchi, S. Sobajimo, S. Murahashi and Y. Matsui, *J. Chem. Phys.* 35, 369 (1961).
[22] G. Zerbi and M. Gussoni, *Spectrochim. Acta*, 22, 2111 (1966).
[23] V. La Garde, H. Prask and S. Trevino, *Disc. Faraday Soc.* 48, 15 (1969).
[24] W. Myers, G. C. Summerfield and J. S. King, *J. Chem. Phys.* 44, 184 (1966).
[25] S. Trevino and H. Boutin, *J. Macromol. Sci. A1*, 723 (1967).
[26] M. Tasumi, T. Shimanouchi and T. Miyazawa, *J. Molec. Spectrosc.* 9, 261 (1962).
[27] T. P. Lin and J. L. Koenig, *J. Molec. Spectrosc.* 9, 228 (1962).
[28] L. Piseri and G. Zerbi, *J. Chem. Phys.* 48, 356 (1968).
[29] G. Zerbi and L. Piseri, *J. Chem. Phys.* 49, 3840 (1968).
[30] L. Piseri and G. Zerbi, *J. Molec. Spectrosc.* 26, 254 (1968).
[31] F. J. Boerio and J. L. Koenig, *J. Chem. Phys.* 52, 4826 (1970).
[32] D. I. Marsh and D. H. Martin, *J. Phys. C, Solid State*, 5, 2309 (1972).
[33] E. B. Wilson, J. C. Decius and P. C. Cross, *Molecular Vibrations*. McGraw-Hill: New York (1953).
[34] B. Fanconi, *J. Chem. Phys.* 57, 2109 (1972).
[35] B. Fanconi, E. W. Small and W. L. Peticolas, *Biopolymers*, 10, 1277 (1971).
[36] R. D. Singh and V. D. Gupta, *Spectrochim. Acta*, 27A, 385 (1971).
[37] A. B. Dempster and G. Zerbi, *J. Chem. Phys.* 54, 3600 (1971).
[38] P. T. T. Wong and E. Whalley, *J. Chem. Phys.* 55, 1830 (1971).
[39] A. Pellegrini, D. R. Ferro and G. Zerbi, *Molec. Phys.*, in press.
[40] J. E. Bertie and E. Whalley, *J. Chem. Phys.* 46, 1271 (1967);
J. E. Bertie and E. Whalley, *J. Chem. Phys.* 40, 1637 (1964).
[41] P. Faure and A. Kahane, *International Conference on Phonon. Rennes*, p 243. Edition Flammarion; Paris (1971).
[42] B. Renker, International Symposium on Physics and Chemistry of Ice, Ottawa (1972).
[43] P. Bosi, R. Tubino and G. Zerbi, International Symposium on Physics and Chemistry of Ice, Ottawa (1972).
[44] P. Bosi, R. Tubino and G. Zerbi, *J. Chem. Phys.*, in press.
[45] P. Torkington, *J. Chem. Phys.* 17, 347 (1949).
[46] P. H. Lindenmayer, *J. Polym. Sci. C1*, 5 (1963);
R. Hosemann, *Polymer, London*, 3, 349 (1962);
B. Wünderlich, *J. Polym. Sci. C1*, 41 (1963).
[47] A. Keller, *Rep. Progr. Phys.* 31, 623 (1968).
[48] See for example: N. G. McCrum, B. E. Read and G. Williams, *Anelastic and Dielectric Effects in Polymeric Solids,* Wiley: New York (1967).
[49] E. W. Fischer, *International Symposium on Macromolecules, Leiden* (*1970*), Butterworth: London (1971).
[50] G. Zerbi, *Pure Appl. Chem.* 26, 499 (1971).
[51] A. Zambelli, C. Tosi and C. Sacchi, *Macromolecules*, 5, 649 (1972).
[52] M. Tasumi and G. Zerbi, *J. Chem. Phys.* 48, 3813 (1968).
[53] G. Zerbi, L. Piseri and F. Cabassi, *Molec. Phys.* 22, 241 (1971).
[54] G. Zerbi, *International Symposium on Phonons, Rennes*, p 248. Edition Flammarion: Paris (1971).
[55] M. Sacchi and G. Zerbi, *Macromolecules*, in press.
[56] G. Masetti, F. Cabassi and G. Zerbi, *Macromolecules*, in press.
[57] I. M. Lifshitz, *Nuovo Cimento* (Suppl. 10), 3, 716 (1956).

[58] P. G. Dawber and R. J. Elliott, *Proc. Roy. Soc.* **A273**, 222 (1963);
P. G. Dawber and R. J. Elliott, *Proc. Phys. Soc.* **81**, 453 (1963).

[59] K. Hölzl and C. Schmid, *J. Phys. C; Solid State Phys.*, **5**, L185 (1972).

[60] See for example: L. Genzel, in *Optical Properties of Solids*, edited by S. Nudelman and S. S. Mitra, Plenum: New York (1962).

[61] A. J. Sievers, Nato Advanced Study on Elementary Excitations and their Interactions, Cortina d'Ampezzo (1966).

[62] P. Dean, *Proc. Roy. Soc.* **A254**, 507 (1960).

[63] P. Dean, *Rev. Mod. Phys.* **44**, 127 (1972).

[64] J. H. Wilkinson, *The Algebraic Eigenvalue Problem*. Clarendon: Oxford (1965).

[65] A. A. Maradudin, E. W. Montroll and G. H. Weiss, *Solid State Physics*, Vol. V. Academic Press: New York (1963).

[66] G. Zerbi, 'Defects in organic crystals; numerical methods', *Enrico Fermi Summer School on Lattice Dynamics and Intermolecular Forces, Varenna* (1972). *Nuovo Cimento,* in press.

[67] M. Gussoni and G. Zerbi, To be published.

[68] T. Miyazawa, *J. Polymer Sci.* **C7**, 59 (1964).

[69] M. J. Gall, P. J. Hendra, C. J. Peacock, M. E. A. Cudby and H. A. Willis, *Spectrochim. Acta,* **28A**, 1485 (1972).

[70] M. J. Gall and P. J. Hendra, *The Spex Speaker,* **16**, No. 1, 1 (1971).

[71] A. Keller, *Kolloid Zh.* **197**, 98 (1964).

[72] M. I. Bank and S. Krimm, *J. Appl. Phys.* **39**, 4951 (1968).

[73] C. Hall, J. W. Fleming, G. W. Chantry and J. A. D. Matthew, *Molec. Phys.* **22**, 325 (1971).

[74] M. Lax and E. Burstein, *Phys. Rev.* **97**, 39 (1955).

[75] H. Schonhorn and J. P. Luongo, *Macromolecules,* **2**, 366 (1969).

[76] C. W. Bunn and E. R. Howells, *Nature, London,* **174**, 549 (1954).

[77] E. S. Clark, *J. Macromol. Sci. Phys.* **B1(41)**, 795 (1967).

[78] P. De Santis, E. Giglio, A. M. Liquori and A. Ripamonti, *J. Polymer Sci.* **A1**, 1383 (1963).

[79] M. J. Hannon, F. J. Boerio and J. L. Koenig, *J. Chem. Phys.* **50**, 2829 (1969).

[80] C. J. Peacock, P. J. Hendra, H. A. Willis and M. E. Cudby, *J. Chem. Soc. (A),* 2943 (1970).

[81] F. J. Boerio and J. L. Koenig, *J. Chem. Phys.* **54**, 3667 (1971).

[82] R. E. Shawyer and P. Dean, *J. Phys. C; Solid State Phys.* **5**, 1028 (1972).

NUCLEAR MAGNETIC RESONANCE OF SYNTHETIC POLYPEPTIDES

E. M. Bradbury, P. D. Cary, C. Crane-Robinson and P. G. Hartman

Biophysics Laboratories, Physics Department, Portsmouth Polytechnic, Gun House, Hampshire Terrace, Portsmouth PO1 2QG, UK

ABSTRACT

High resolution n.m.r. spectroscopy of synthetic polypeptides has already been extensively applied to a study of conformations and conformational transitions, particularly the helix–coil transition. Since every atom of an amino acid residue can, in principle, be made to yield a spectrum, the potential of the method for investigating the finer details of conformation far exceeds that of optical methods.

The main chain proton resonances have been shown to be sensitive not only to the helix content (and helix sense in the case of poly aspartate esters) but also to the polydispersity of the sample, giving a direct indication of the conformational heterogeneity at all points through the transition. Earlier suggestions that multiple αCH and NH peaks were the reflection of a slow step in the conversion of a helical residue into a solvated coil state, have now been shown to be unfounded. The αCH shift difference between the two conformational states has been found not to depend on the nature of the polypeptide, but rather to be a function of the solvent system used; in all cases so far observed however, the helical αCH was upfield of the coil. In a situation for which ORD/CD data were ambiguous (poly L tyrosine) this upfield displacement of the peak was itself taken as indicating a coil to helix transition. The same criterion for the existence of a transition has also been applied in the study of the conformations of racemic DL copolypeptides and the method shown to be particularly appropriate for these polymers. The main chain resonances of poly L alanine have long been the subject of discussion, due both to the presence of unusual additional peaks (that are now seen to be in part an end effect) and to the fact that multiple αCH resonances are not observed through the helix–coil transition. It has been shown that this is a direct consequence of the low cooperativity of transition in this case, and other instances of related behaviour have been detailed.

The ability of n.m.r. to study the conformation of the sidechain in addition to that of the backbone, has been exploited the most fully for poly β benzyl L aspartate. This polymer was chosen since it has been calculated that the left handed helix sense is a consequence of specific interactions that involve a strongly preferred sidechain conformation. Analysis of the vicinal and gem spin-coupling led to the result that no such rigidity exists in the sidechain and therefore that a reappraisal is required of the interactions leading to this unusual helix sense.

In AB copolymers, whether block or random, n.m.r. can be used to follow the behaviour of the A and the B components separately. This has been illustrated with a number of glutamate–aspartate copolymers and led to conclusions regarding both helix sense and also the details of the copolymerization process not readily available by other methods.

53

It may be stated in conclusion that the diversity of conformational problems that have been tackled by high resolution n.m.r. serves to indicate that the method is not merely a supplement to the well known optical techniques but a powerful tool in its own right.

INTRODUCTION

Although much simpler than proteins, synthetic polypeptides have served as model compounds for a number of conformational features found in proteins. The main approach has been to study synthetic polypeptides by various techniques and establish the physical and optical parameters which are characteristic of regular and defined conformations of the polypeptide chain. Thus ORD/CD and also infra-red spectroscopy have been very successful in defining spectral properties of the α helical and regular extended β conformations in solution and in films. These techniques, however, provide information on the backbone conformation alone and even these data may be unobtainable in ORD/CD if sidechain chromophores obscure those of the main chain. High resolution n.m.r. spectroscopy has the greatest potential of all of the spectroscopic techniques for conformational studies in solution since many different nuclei in a molecule can be studied separately, especially at high fields, and in particular the sidechain spectrum is normally quite separate from that of the main chain. Moreover, each resonance is characterized by three parameters, the shift, multiplicity and relaxation times, all of which may be conformationally dependent. A second advantage of n.m.r. lies in the fact that the time scale of the technique can result in the spectrum being dependent on molecular motions, in particular conformational transitions, whereas the optical spectroscopic techniques always yield a snapshot picture, with each conformation making its own contribution. No information on the molecular dynamics is normally obtainable. This vast potential of n.m.r. is only at the beginning of its realization in biological studies and it is important to lay the ground work for proteins as soundly as possible by studies of synthetic polypeptides.

This paper makes no attempt to review the field but rather to highlight issues of greater importance that can be studied by n.m.r. (such as the details of the helix–coil transition), and also to discuss polymers having unusual or uncertain conformational properties such as poly tyrosine, poly alanine and poly aspartates. Correlation of n.m.r. data with those for the same system studied by ORD/CD (or by infra-red) is important in this work and the ORD parameter used is b_0 [$\sim 0°$ for random coil $\pm \sim 630°$ for a left handed (LH) and right handed (RH) helix].

THE HELIX–COIL TRANSITION

Figure 1 shows the temperature induced helix–coil transition of that well-known and fairly well-behaved polypeptide poly γ benzyl L glutamate (PBLG) in chloroform/TFA[1]. The apparent linewidths in the random coil form (of the order of 15 Hz for the main-chain αCH) are greater than that of small molecules in this solvent system, but not too great to preclude study. It was early established[2] that segmental motion of the chain was responsible for these relatively sharp lines. The true linewidth of the αCH resonance may be about 5 Hz, the remainder being a consequence of amide and β proton coupling. The linewidths of the helical form in *Figure 1* are seen to be similar

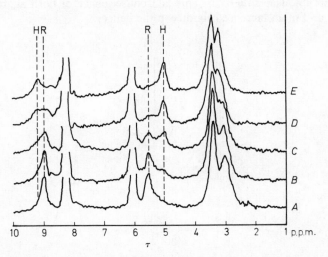

Figure 1. Proton magnetic resonance spectra at 100 MHz showing the temperature induced helix–coil transition of poly γ benzyl L glutamate R10, DP = 92, in 8% TFA–92% $CDCl_2$. H—helix, R—random. A, 0°C, $b_0 = 0$; B, 6°C, $b_0 = -100°$; C, 12°C, $b_0 = -220°$; D, 25°C, $b_0 = -340°$; E, 50°C, $b_0 = -470°$. Peak assignments in A: 7.8 p.p.m., amide NH; 7.4 p.p.m., aromatics; 5.1 p.p.m., benzyl CH_2; 4.6 p.p.m., αCH; 2.5 p.p.m., γCH_2; 2.1 p.p.m., βCH_2.

to those of the coil, which is unexpected. This is not a consequence of the increased temperature used to promote the helix form, since decreased TFA concentration at constant temperature has a similar result. The helical form in *Figure 1* cannot therefore be a rigid rod and the relatively sharp lines (in particular the αCH proton) must be a result of the presence of a small percentage of coil. Such an interrupted helix would be more compact than a rigid rod, having a more rapid overall correlation time and greater internal flexibility. Furthermore, helix–coil interconversion is thought to be a fast process[3] and the rapid motion of a small coil section along the largely helical molecule could result in linewidths not markedly different from those of the coil. Although this matter will be discussed further, particularly as regards αCH linewidths, it may be stated that a full understanding is not yet possible of the motions responsible for the relatively sharp lines of helical polypeptides in TFA-containing solvents. In pure water, however, all polymers so far studied[4, 5, 6] show a distinct broadening of the αCH resonance as the helicity rises and it has not been possible to observe the αCH peak of for example fully helical poly L glutamic acid or poly L lysine. The greater linewidths observed in water are probably a consequence of intermolecular aggregation that does not take place in most organic solvents. With increasing molecular weight in a defined solvent there is an increase in linewidth of several peaks to a limiting

55

value and this can be used to estimate molecular weight[7]. Moreover, in both helix and coil, linewidths decrease for all polypeptides as the distance of the proton from the main chain increases, reflecting an increasing molecular flexibility[8]. In pure chloroform linewidths are much greater than in haloacetic acid containing solvents with the amide NH and αCH resonances often being unobservable (see *Figure 2*). This is a consequence of both aggregation of helices and of an increased rigidity in the helices.

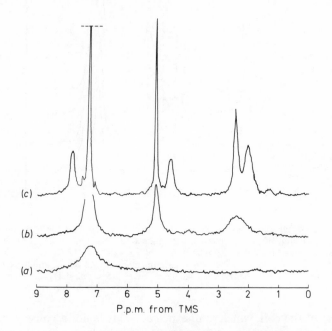

Figure 2. Proton magnetic resonance spectra at 100 MHz of PBLG in solution in (a) $CDCl_3$, (b) 2% TFA–98% $CDCl_3$, and (c) 20% TFA–80% $CDCl_3$.

Figure 1 shows three principal changes in the spectrum as the coil is converted into helix:
(1) the βCH_2 moves downfield by about 0.3 p.p.m. (see ref. 44, Figure 5),
(2) the αCH moves upfield by about 0.5 p.p.m. and
(3) the amide NH moves downfield by about 0.2 p.p.m.
This so-called 'double-peak' phenomenon over the helix–coil transition for the main chain amide and α protons[9] has been the subject of much debate and the position will be reviewed here, largely in terms of the αCH resonance.

THE 'DOUBLE-PEAK' CONTROVERSY

For poly γ benzyl L glutamate (PBLG) in chloroform ($CDCl_3$)–trifluoro acetic acid (TFA) the observation of individual helix and coil αCH peaks at the mid-transition point separated by about 50 Hz (at 100 MHz) suggests that the lifetime of a residue in both conformations is in excess of 3×10^{-1} sec. However, both theoretical[3a] and experimental studies by dielectric relaxation[10] and temperature-jump[3b] techniques indicate that helix–coil interconversion is at least four orders of magnitude faster than this. Such a rapid rate would result in an averaging of both the αCH chemical shift and the peak width over the time a residue spends in the helix and coil states. On the assumption that this averaging is the same over the whole chain, a 'single shifting peak' would thus be observed for the αCH resonance as coil is converted into helix. High molecular weight PBLG in $CDCl_3$–TFA does indeed show such a 'single shifting peak' as does poly L alanine (which will be discussed in detail later) but the majority of polypeptides so far studied in various solvent systems including $CDCl_3$–TFA show the 'double peak' αCH spectrum. Examples of other polymers studied in $CDCl_3$–TFA are poly β methyl L aspartate (PMLA) and poly L leucine[9], poly β benzyl L aspartate (PBLA)[4, 11, 12], poly L methionine[4, 7, 13, 55] and poly L phenylalanine[14]. 'Double peak' behaviour has also been observed in other solvent systems; poly L arginine in methanol–water[15], poly L tyrosine (PLT)[16] in water–dimethyl sulphoxide (DMSO) and PBLG and PBLA in $CDCl_3$–DMSO[17].

This assignment of the two αCH peaks to helix and coil rests on the correlation of the two peak areas with the ORD parameter b_0 over the helix–coil transition; this correlation was found to be excellent for PBLG in $CDCl_3$–TFA[1] and is shown in *Figure 3*. A good correlation was also established[13]

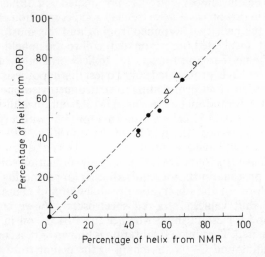

Figure 3. Correlation of n.m.r. estimates of helix content (from αCH intensities) with ORD estimates (from b_0) for three samples of PBLG of DP = 92 (O), DP = 21 (△), and DP = 13 (●).

for poly L methionine at both 60 and 220 MHz using change of TFA concentration rather than temperature. The actual assignment of the two αCH components has, however, been challenged by Joubert *et al.*[6] who proposed that the upfield ('helix') peak is due to unsolvated coil residues whilst the lowfield ('coil') peak is due to solvated coil residues. The shift difference between the peaks is thus attributed entirely to interaction with solvent in the coil state and solvation is postulated to be a slow process since the solvated and unsolvated states are magnetically distinct. The helical residues were postulated to give rise to resonances too broad to be observed. This interpretation has been investigated[18] for PBLG in CDCl$_3$–TFA over the complete helix–coil transition region by measurement of the summed area of both peaks relative to internal standards. The total αCH peak area varied by only eight per cent over the complete transition and it is therefore clear that the αCH resonance of helical residues is almost fully observable under these conditions of measurement.

A second theory of the 'double-peak' phenomenon has been advanced by J. H. Bradbury and co-workers[19] and supported by Tam and Klotz[20]. This theory also involves a slow solvation step as the essential process by which two magnetically distinct αCH proton states are observed. In this case solvation is regarded as complete protonation of the amide group by the acid. The upfield peak is regarded as a composite of unsolvated helix and coil in rapid equilibrium (to accord with the kinetic results) and the low field peak is composed of *protonated* helix and coil residues—also in rapid equilibrium. The shift difference between the two peaks is thus regarded as entirely due to solvation, there being no intrinsic dependence of shift on conformation. Inasmuch as a helix could never maintain a high charge the low-field peak must therefore be due almost entirely to protonated coil. It has been pointed out above that in certain cases an extremely good correlation of peak areas is obtained with the helicity determined from b_0 and this must mean that the contribution of unsolvated (unprotonated) coil to the upfield peak must be slight. This scheme therefore reduces to: upfield-peak unprotonated helix, downfield—completely protonated coil. To test this hypothesis, polypeptides have been studied in a non-protonating solvent, dimethylsulphoxide (DMSO) as follows. PBLA is random coil in pure DMSO and on addition of chloroform takes up the usual LH helical conformation[17, 27] with a typical 'double peak' being observed near the middle of the transition. PBLG samples of low molecular weight have also shown separate helix and coil peaks in pure DMSO (see *Figure 10*). Since DMSO is a non-protonating solvent it can be concluded that protonation is not required either for promoting the helix–coil transition itself, or for the observation of different and characteristic helix and coil αCH shift values. Infra-red spectroscopy, however, is the most powerful method for establishing whether the PBLG coil in CDCl$_3$–TFA is highly protonated. The amide II vibration is known to be a composite of in-plane NH bending and C—N stretching in the planar *trans* amide group. Protonation on either the nitrogen or oxygen atom would result in disruption of this coupling and complete loss of absorption at ~ 1550 cm^{-1}. *Figure 4* shows the 6 μ region of PBLG in CDCl$_3$ as TFA is added to induce transition to the coil at about 12 per cent acid. At 15 per cent TFA both amide I (at ~ 1660 cm^{-1}) and amide II (at ~ 1550 cm^{-1}) are somewhat broader than in the helix, as expected, but little changed in frequency. There cannot therefore be a large amount of protonation of the amide groups. Solvation of the sidechain ester carboxyl (at ~ 1735 cm^{-1}) by hydrogen bonding is apparent, however.

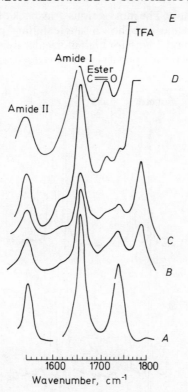

Figure 4. Infra-red spectra of 0.2M solutions of poly β benzyl L glutamate in (A) CDCl$_3$; (B) 2.5% TFA–97.5% CDCl$_3$; (C) 5% TFA–95% CDCl$_3$; (D) 10% TFA–90% CDCl$_3$; (E) 15% TFA–85%CDCl$_3$.

A third theory of the 'double peak' phenomenon has been advanced by Ferretti and co-workers[21] who propose helix nucleation as the essential slow step in the kinetic scheme. The very much faster helix and coil propagation rates along the chain are taken by these authors to determine the kinetic times of the order of $\sim 10^{-6}$ sec whilst the longer n.m.r. lifetimes that give rise to separated peaks are calculated to be a consequence of the much slower helix nucleation. The predicted line shape patterns vary with molecular weight and with the assumptions made as to the different nucleation and propagation rate constants. The αCH line shapes are characterized by a purely 'random coil' peak that remains fixed in shift and an average helix peak that moves towards the coil peak for all except the lowest molecular weights. Spectra of PBLG of intermediate molecular weights (DPs from 160 to 300) appear to show such an effect[22]. We have studied a very large number of PBLG samples of different molecular weights and find that for samples of DP between 150 and 400 the midpoint of the transition is characterized by a broad resonance having a maximum closer to the coil position for some samples and closer to the helix position for others. With increase of molecular weight the maximum is found to move to a shift midway between the extremes and finally spectra are observed similar to those shown in *Figure 5* for a high molecular weight sample. The most serious drawback of the theory, however, is that it offers no explanation of how a 'single shifting' peak can be generated and this has been observed both for high molecular weight samples of **PBLG** and for samples of low polydispersity (see below, *Figures 5* and *9* respectively).

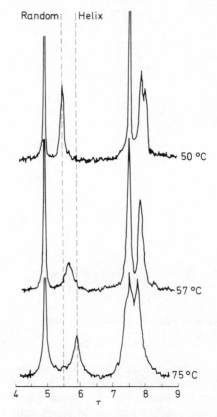

Figure 5. 220 MHz proton spectrum of PBLG, DP = 640, in 25 % TFA– 75 % CDCl$_3$.

Ullman[23] has given a theory of the 'double peak' phenomenon that is the most satisfactory to date since it offers an explanation which is quite independent of polymer or solvent system and predicts in general terms the observed dependence of the αCH peak shapes on the molecular weight of the sample. Ullman's theory assumes that there is rapid interconversion of helix and coil states. Two factors could be responsible for the appearance of the 'double peak' according to this approach: first, the free energy difference between helix initiation and helix continuation means that in the helix–coil transition region a residue near the end of a polypeptide chain has a greater probability of being in the coil state than a residue in the middle of the chain, i.e. averaging of helix and coil states is not equivalent for all residues in the chain, resulting in asymmetric peaks. Although this effect would disappear

60

for very high molecular weight, an averaged and shifting αCH peak resulting (as is in fact observed), it is not sufficient to explain the occurrence for low molecular weights of quite separate helix and coil peaks that remain fixed in position through the transition (see *Figure 1*). The second contributing factor (as suggested, independently, by Jardetzky[24, 25]) is molecular weight polydispersity. Under specified solvent conditions the helicity of a given polypeptide chain is strongly dependent on the molecular weight, if that is low. A low molecular weight polydiperse sample therefore in the middle of the helix–coil transition will consist in the main of molecules that are either largely helical (and so contribute to the upfield peak) or largely coil (and so contribute to the lowfield peak). On this basis a good correlation of the upfield 'helix' peak area with b_0 (as seen in *Figure 3*) would be observed only for samples having a molecular weight spread broad enough such that under any conditions only a small proportion of the molecules are actually in the process of the helix–coil transition. These postulates have been subjected to an experimental test by the study of several PBLG samples in $CDCl_3$–TFA[26]. The sample of *Figure 6*(a) (R10) having an average DP of 92, was found to have the same M_w and M_n, within experimental error, as sample S416 whose spectra are shown in *Figure 6*. A sample of DP ~ 270

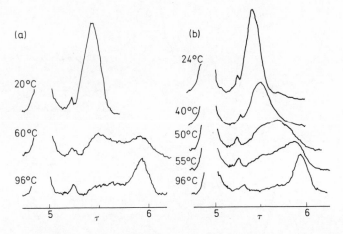

Figure 6. 100 MHz proton spectra, αCH region, of (a) PBLG, DP = 92 (R10), and (b) PBLG, DP = 100 (S416), both in 20% TFA–80% $CDCl_3$.

(No. 314) gave αCH line shapes at 100 MHz intermediate between those of *Figures 5* and *6*(a) and not differing greatly from those of S416 in *Figure 6*(b). Moreover, the αCH line shapes of PBLG 314 remained virtually unchanged on remeasurement at 220 MHz. This line shape intermediate between the double peaks of *Figure 1* and the 'single shifting' peak of *Figure 5* therefore represents a continuous range of shift values, i.e. of helicities, within the sample, over the transition region. The high molecular weight sample SP 18.4 (DP ~ 640) (*Figure 5*) shows what approximates to a 'single shifting peak' in the transition region, although careful measurement shows that the linewidth rises to a maximum near the middle of the transition. Such changes in linewidth have also been noted for poly L methionine[4]. If polydispersity were the cause of the 'double-peak' phenomenon such that increasing molecular weight results in a change of the line shapes from double peaks to single in the transition region, why then should R10 and S416 give markedly different spectra? Gel permeation chromatography was used to demonstrate that R10 is much more polydisperse than S416 and R10 is therefore very polydisperse in helicity in the transition region. Fractionation of R10 by means of precipitation chromatography has yielded components having weight averages between 20 and 190. *Figure 7* compares the αCH spectra of fractions from R10 having weight averages 170 and 45, which ORD measurements

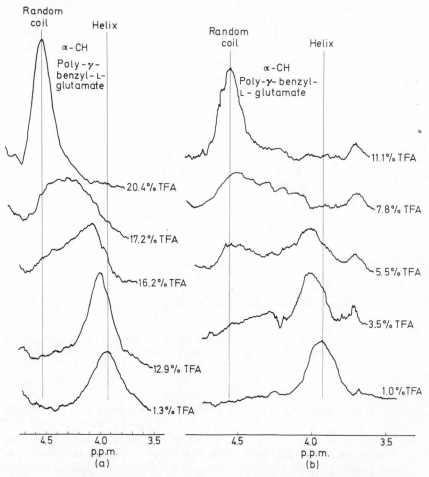

Figure 7(a). 100 MHz spectra in TFA–CDCl$_3$ at 22°C of sample (170 mer) fractionated from R10 ($\overline{DP}_w = 92$).

Figure 7(b). 100 MHz spectra in TFA–CDCl$_3$ at 22°C of sample (45 mer) fractionated from R10 ($\overline{DP}_w = 92$).

show to have helix–coil transition midpoints at 65 per cent and 40 per cent DCA respectively, i.e. very well separated. That polydispersity is a major cause of the 'double-peak' phenomenon in R10 is thus established. Nevertheless, the 45 mer of *Figure 7* still shows a separation of the helix and coil peaks.

Very recently Nagayama and Wada[27] subjected a sample of PBLG to gel permeation chromatography using DMF as solvent and obtained a fraction having $\overline{DP}_n = 43$. *Figure 8* shows its molecular weight distribution obtained

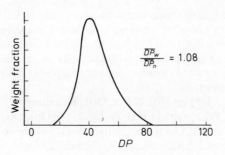

Figure 8. Molecular weight distribution of PBLG sample fractionated by GPC in DMF solution.

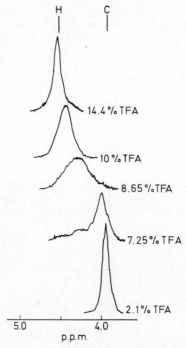

Figure 9. 220 MHz αCH spectrum of fractionated PBLG sample in the helix–coil transition region in TFA–CDCl$_3$. Polymer concentration, about 3% w/v. H denotes helix position, C, coil position.

by rechromatography. Despite this low molecular weight the appearance of the αCH resonance through the helix coil transition ($CDCl_3$–TFA) was essentially that of a single shifting peak (see *Figure 9*). The authors conclude that polydispersity and end effects are the sole cause of the 'double-peak' phenomenon and therefore that no discrepancy whatever exists between the n.m.r. line shapes and kinetic measurements in the helix–coil transition region.

THE CAUSES OF THE HELIX AND COIL SHIFT-DIFFERENCE

Several solvents have been used to study the helix–coil transition of polypeptides. Data on αCH chemical shifts are presented in *Table 1* for those cases in which both fully helical and fully coiled states have been observed. Small errors could, however, be present in certain of the shift values as a consequence of some conformational impurity. Certain data require explanatory comment.

The random coil data on PBLG in $CDCl_3$ and dimethylformamide (DMF) and on PBLA in $CDCl_3$ are indirect, being derived from DL polymers (see also below). The data for PBLG and PBLA in pure DMSO have been obtained in a recent study[17,28] of homo and copolypeptides. *Figure 10* shows the αCH

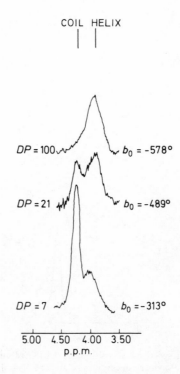

Figure 10. 100 MHz spectra, αCH region, of three samples of PBLG in $DMSOd_6$ at 30°C.

peak of PBLG samples in DMSO over a considerable molecular weight range and it can be seen that whilst long chains are helical in this solvent, short chains have a large coil component. Pure PBLA is random coil in DMSO[56] and so are random copolymers with PBDG up to 50 per cent

Table 1

Polymer	Solvent at transition midpoint	αCH Helix shift (p.p.m.)	αCH Coil shift (p.p.m.)	αCH $-\Delta_{H/C}$ (Helix–coil shift difference) p.p.m.	Ref.
PBLG	CDCl$_3$–14%TFA	3.95	4.45	0.50	1, 4
PBLG	Pure CDCl$_3$	3.95	4.2–4.3	0.35–0.25	49
PBLG	27%D$_2$O–TFA	4.20	4.70	0.50	53
PBLG	41% formic acid–TFA	4.25	4.70	0.45	53
PBLA	CDCl$_3$–1.5%TFA	4.30 (LH)	4.80	0.50	11, 43
		4.40 (RH)	4·80	0.40	11, 43
PBLA	Pure CDCl$_3$	4.30 (LH)	4.3	0.0	49
Poly D α-amino n-butyric acid	CDCl$_3$–15%TFA	3.80	4.35	0.55	32
PCLL	CDCl$_3$–9%TFA	3.90	4.45	0.55	54
Poly L leucine	CDCl$_3$–30%TFA	4.08	4.55	0.47	32, 36
Poly L methionine	CDCl$_3$–40%TFA	4.23	4.75	0.52	4, 13
Poly L phenyl alanine	CDCl$_3$–5%TFA	4.20	4.70	0.50	14
PLT	10%D$_2$O–DMSO	4.10	4.44	0.33	16
Poly L arginine	85%MeOH–D$_2$O	4.0	4.4	0.4	15
PBLG	Pure DMSO	3.95	4.26	0.31	17
PBLA	Pure DMSO	4.30 (LH)	4.64	0.34	17, 28
PBLG	DMF	4.09	4.39	0.30	11, 49
PGA	D$_2$O/pD 4.8	4.2	4.3	0.1	4, 5
PLL	D$_2$O, pD 10.4	4.2	4.3	0.1	5
Poly–N^5-(4-hydroxy-butyl) L glutamate	D$_2$O (20°C)	4.10	4.26	0.16	6
Copoly L glutamic acid[42], L lysine HBr[28], L alanine[30]	D$_2$O(20°C)	4.2	4.3	0.1	11

glutamate. Further addition of PBDG causes the polymer gradually to assume the LH helical form. This establishes the LH helical shift of PBLA as 4.3 p.p.m. (similar to that in CDCl$_3$) and the coil shift as 4.69 p.p.m. The RH helical form of PBLA has not yet been observed in pure DMSO, although in DMSO–60 per cent CDCl$_3$ cosolvent the αCH shift is 4.40 p.p.m., as in pure CDCl$_3$.

65

The data in *Table 1* show that although the αCH helix position is always upfield of the coil, the magnitude of this difference $\varDelta_{H/C}$ is very dependent on the solvent system, varying from 0.0 to 0.5 p.p.m. The general conclusion to be drawn from *Table 1* is clearly that $\varDelta_{H/C}$ values are characteristic of a particular solvent system, rather than of the polymer. This suggests that $\varDelta_{H/C}$ largely results from solvation differences between the helix and the coil, rather than being a difference intrinsic between the two conformations.

WATER SOLUBLE POLYPEPTIDES

Table 1 shows that in water the dependence of the αCH shift on helix content is lower than in organic solvent systems. The peak displacement has been followed for both poly L glutamic-acid (PGA) and poly L lysine (PLL) over the helix–coil transition[4, 5], but in both cases aggregation of helical polymer with consequent line broadening prevented observation of the fully helical state. The value of $\varDelta_{H/C}$ for these polymers was indicated to be not less than 0.1 p.p.m. A copolymer of molar composition (L-glutamic acid[42], L-lysine HBr[28], L-alanine[30]) has also been studied in water[5], and the helicity varied by both pH and temperature change. Only a small upfield shift of the αCH peak was noted on helix formation amounting (on measurement at 220 MHz[11]) to 0.04 p.p.m. as the helicity changed from 32 to 71 per cent. This implies a value of ~ 0.1 p.p.m. for the complete transition. Joubert *et al.*[6] have studied several glutamine derivatives in water the helicity of which is dependent on temperature. This allowed the helix–coil transition to be studied without any complicating effects due to sidechain ionizations, though aggregation of helical polymer was just as evident as with PGA or PLL. In the case of poly N-(hydroxybutyl) L glutamine (PHBG) at 20°C a shift of 4.26 p.p.m. was observed for the coil form and 4.13 p.p.m. for a helix content of 67 per cent. The extrapolated value of $\varDelta_{H/C}$ for the complete transition of PHBG was 0.16 p.p.m. In all the above studies in water the αCH resonance appeared to move as a 'single shifting peak' and did not exhibit any obvious 'double-peak' character. Since, however, the displacements are of the same order as the linewidths and aggregation broadening takes place on helix formation, 'double peak' behaviour would be largely obscured.

The above results show that for protein conformational studies in water the αCH shift is not likely to be of great value as a consequence of its low dependence on helicity.

Poly L alanine

The spectrum of poly L alanine (PLA) in $CDCl_3$–TFA–DCA has been the subject of much discussion in the literature for two reasons. First, as seen from *Figure 11*, the principal αCH resonance has the appearance of a 'single-shifting peak' even for samples of low molecular weight[4, 29, 30, 31]. This behaviour is normally associated only with polymers of the highest molecular weight (e.g. see *Figure 5*). The PLA peptide NH resonance remains approximately constant in shift over the same range of solvent composition from 30 to 100 per cent TFA, moving only ~ 0.08 p.p.m. upfield. This approximate constancy of the NH peak is probably the result of a balance between upfield displacement by ~ 0.2 p.p.m. (the value observed for PBLG in *Figure*

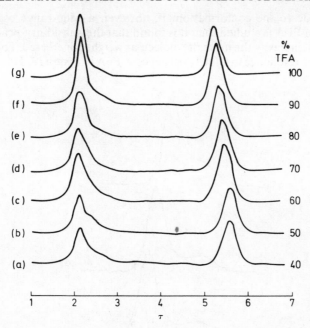

Figure 11. 60 MHz spectra of poly L alanine in TFA–CDCl$_3$. Amide NH and αCH protons.

1) due to the breakdown of the helical conformation and a downfield displacement of the peak by about the same amount due to the overall increase of TFA, as discussed above for the αCH proton. The small net displacement of the NH peak nevertheless shows the same form as the change in b_0. Over the same solvent composition the single αCH resonance moves ~ 0.30 p.p.m. downfield. This too correlates with the b_0 changes, although the magnitude of the displacement is in this case augmented by the bulk TFA effect. Thus although the PLA transition is not sharp and differs thereby from PBLG, there is no reason on the basis of the above to doubt that the principal αCH and NH peaks of PLA represent an average of helix and coil states and are therefore a good indication of conformation, in the same way as for high molecular weight PBLG. The reasons underlying the observation of a 'single shifting' αCH peak for PLA rather than the more usual 'double-peaks' will be discussed later.

Detailed study of PLA[31, 32] reveals, however, that there is a subsidiary NH peak (at ~ 7.7 p.p.m. in 80 per cent TFA–CDCl$_3$) and a subsidiary αCH peak (at ~ 4.7 p.p.m. in 80 per cent TFA–CDCl$_3$). These peaks are denoted by the letter S in *Figures 12* and *13*. Both these peaks move to lowfield by 0.1–0.2 p.p.m. over the TFA range of 30 to 100 per cent as can be seen from *Figure 12*. Their shifts are just those found for the NH and αCH peaks of poly DL alanine[30] in CDCl$_3$–TFA and their areas grow slightly at the expense of the main peaks with increasing TFA. For these reasons Ferretti and Paolillo[31] assigned the subsidiary peaks to random coil, and the main peaks to fully helical polymer thereby concluding that PLA is largely helical in TFA and b_0

67

is no guide to the conformation. If, however, a wide range of molecular weights of PLA is studied then it is found that the subsidiary peaks become reduced in intensity the more the molecular weight is increased. This is shown for the N*H* resonance in $CDCl_3$–80 per cent TFA in *Figure 13*. This figure also

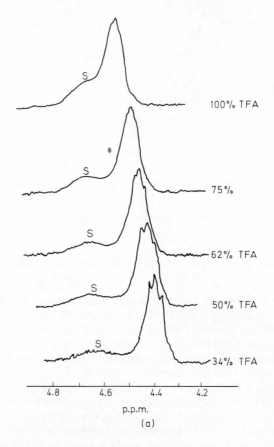

Figure 12(a). 220 MHz spectra, α*CH* region, of poly D alanine, $\eta_{sp}/c = 9.04$ dl mol^{-1}, in TFA–$CDCl_3$.

shows the coincidence of the subsidiary peak shift with that of poly DL alanine. This dependence on molecular weight and the correspondence of the shift with that of poly DL alanine suggests that the subsidiary peaks can be assigned to PLA in some coil structure taken up by low molecular weights and/or ends of chains.

If these interpretations of the PLA peaks are correct it implies a coexistence of two states: (i) virtually pure 'random coil', and (ii) rapidly interconverting helix and 'coil', each of which gives rise to its own distinct peak. The reasons underlying this apparent separation of states may be just those given above

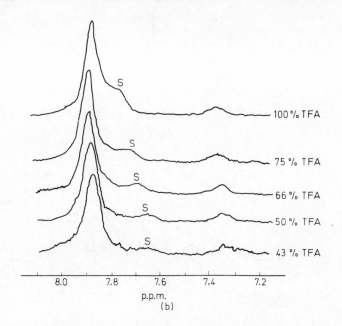

Figure 12(b). 220 MHz spectra, N*H* region, of poly D alanine, $\eta_{sp}/c = 9.04$ dl mol^{-1}, in TFA–CDCl$_3$.

for the 'double-peak' α*CH* resonance of PBLG in that there can be two contributions. The first factor is polydispersity, as a result of which the two states exist in separate molecules (low molecular weights favouring the highly 'coiled' state, and higher molecular weights the helix–coil inter-conversion state). In the absence of any fractionation of PLA we cannot yet assess this contribution. The second factor is end effects in that residues near the ends of the chains have an enhanced probability of being in the purely 'coiled' state. We have investigated[33] this factor by studying B—A—B co-polymers, where A represents poly L alanine and B represents PBLG, PBLA or poly ε-carbobenzoxy L lysine (PCLL), in CDCl$_3$ from 0 to 100 per cent TFA. The important results of this study were:

(1) that the α*CH* peak moved downfield as a single shifting peak from ∼4.1 p.p.m. in CDCl$_3$ to ∼4.6 p.p.m. in 100 per cent TFA. These are shift values close to those observed for the helix–coil transition of PBLG in CDCl$_3$–TFA. *Figure 14* shows expanded spectra for a PBLG block copolymer with PLA which illustrate this point.

(2) The α*CH* shift paralleled closely the changes in b_0. *Figure 15* shows a decomposition of b_0 data for a PBLG block copolymer with PLA and the corresponding α*CH* shift data.

(3) Despite the fact that the PLA block was of low molecular weight there was no sign of the 'subsidiary' α*CH* or N*H* peaks. This is most clearly demonstrated for the α*CH* peak in spectra of PBLA block copolymers since the PBLA α*CH* is well removed from that of PLA. Since the PLA in these

69

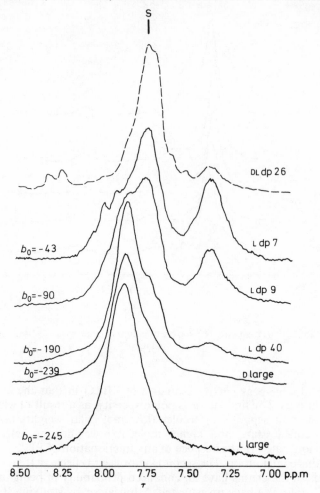

S

DL dp 26

$b_0 = -43$

L dp 7

$b_0 = -90$

L dp 9

$b_0 = -190$

L dp 40

$b_0 = -239$

D large

$b_0 = -245$

L large

8.50   8.25   8.00   7.75   7.50   7.25   7.00 p.p.m

τ

Figure 13. 100 MHz N*H* resonance of poly alanine in 80 % TFA–20 % CDCl₃. Different molecular weights and helix contents indicated.

block copolymers does not have free ends this indicates that the 'subsidiary' peaks in the homopolymer spectrum are largely due to end effects, since in the block copolymer there is little tendency for the L alanine residues near the ends of the PLA block to be predominantly in a 'coil' form. The rapid interconversion of helix and coil then covers the whole of the PLA block. One could also postulate that what is here termed the 'purely coil state' (characteristic also of poly DL alanine) is in fact a special coil structure that exists for times long on the n.m.r. scale due to a slow kinetic step in its formation. There is no independent evidence, however, for a special structure in low molecular weight PLA or in poly DL alanine.

That poly L alanine (and other polypeptides with hydrocarbon sidechains) interact with TFA in an unusual manner has been known for some while[29]

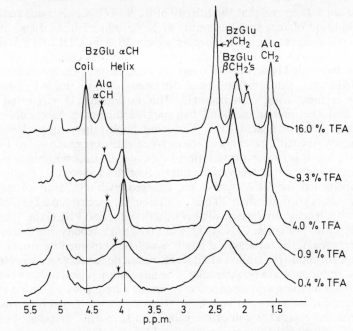

Figure 14. 220 MHz spectra in TFA–CDCl$_3$ of block copoly [benzyl L glutamate (39)–L alanine (46)–benzyl L glutamate (33)]. L Alanine αCH arrowed.

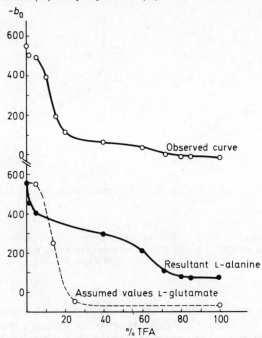

Figure 15. b_0 values in TFA–CDCl$_3$ for block copoly [benzyl L glutamate (39)–L alanine (46)–benzyl L glutamate (33)].

71

and *Figure 15* shows that the helicity of PLA in $CDCl_3$ is much reduced by the addition of very small amounts of TFA, unlike PBLG. In addition the infra-red spectra of high molecular weight PLA in $CDCl_3$–TFA show an unusual band at ~ 1616 cm^{-1} that grows as the TFA content is increased[29]. Since the amide I band at ~ 1655 cm^{-1} simultaneously decreases in intensity, the 1616 cm^{-1} ('interaction') band can be assigned to amide I vibrations of polymer interacting with the acid. This assignment is supported by the fact that the interaction band has parallel dichroism when observed in oriented films of helical PLA. The band might alternatively be assigned to the asymmetric stretching of the charged carboxyl group of the trifluoroacetate ion if polymer protonation takes place. However, it has been observed[34] that the frequency of the interaction band remains constant if DCA is substituted for TFA, whilst the charged carboxyl band of these two acids differs by more than 20 cm^{-1}. Similar i.r. spectra have been observed for poly L leucine. In the case of poly L methionine and PBLG the 'interaction band' is also present in $CDCl_3$–TFA but is of low intensity. Furthermore, the appearance of the 'interaction band' is not accompanied by corresponding changes in amide II. It follows that there cannot be any large scale protonation of the amide group in polypeptides. The interaction is probably one of strong hydrogen bonding and a reduction in amide I frequency would be expected in these circumstances. It is difficult to establish whether this interaction is preferentially with the coil or the helix. The 1616 cm^{-1} band is observed in films of helical poly alanine exposed to TFA or DCA vapour and moreover

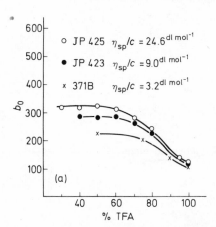

Figure 16(a). b_0 values in TFA–CDCl$_3$ for three poly D alanine samples of widely differing molecular weights.

Figure 16(b). b_0 values in TFA–CDCl$_3$ for two samples of poly D α amino n butyric acid of widely differing molecular weights.

x-ray diffraction studies of such films[35] indicate order in the acid molecules and an overall increase in crystallinity with no apparent loss in polymer helicity. A structure has been proposed for the PLA α helix–acid complex[35]. Thus strong interaction of TFA with helical PLA is probably the cause of the sharp changes in the PLA spectra (n.m.r. and i.r.) as small quantities of TFA are added to a $CDCl_3$ solution (*Figure 14*). In the random coil form of PLA the TFA probably also hydrogen bonds to the amide group since a broad amide I band results, giving a spectrum very similar to that of PBLG at high acid concentrations.

The 'single shifting' αCH peak for PLA (e.g. *Figure 11*) is in strong contrast to the 'double peak' observations for PBLG of similar molecular weight (e.g. *Figure 1*). We have compared by ORD the helix–coil transition of several PLA samples of widely differing molecular weights in the same way as described above for PBLG samples. There are unfortunately no molecular weight calibration data available that are appropriate for PLA and the polymers are characterized therefore only by their viscosities in DCA. Sample 371B [*Figure 16*(a)] is of very low molecular weight. The striking result is that there is very little dependence of helicity on molecular weight for poly alanine, in strong contrast to PBLG. Nowhere over the accessible solvent range does the b_0 differ by more than 50° for the two JP samples. The absence of any marked dependence of helicity on molecular weight is expected for a polypeptide showing a very broad helix–coil transition, i.e. one of low cooperativity and having only short helical segments. This means that in a polydisperse PLA sample there is only a narrow range of helicities present for a given solvent condition and therefore that rapid helix–coil interconversion gives rise to an essentially single peak.

OTHER POLYPEPTIDES WITH HYDROCARBON SIDECHAINS

Poly D α-amino *n*-butyric acid has an ethyl sidechain and *Figure 16*(b) shows b_0 data for two samples of different molecular weight over the helix–coil transition. The molecular weight dependence of the transition is greater than for PLA but still much less than for PBLG and this is reflected in the n.m.r. spectrum shown in *Figure 17* for sample 372. Although the αCH resonance shows apparent multiplicity over the transition, the overall behaviour is intermediate between that of the 'single shifting peak' observed for PLA and that of the 'double peaks' seen in *Figure 1* for PBLG. This is emphasized by the fact that at the transition midpoint (~ 11 per cent TFA from the b_0 data) the αCH resonance does not look symmetrical: the centre of gravity of the resonance, however, does lie at the exact midpoint of the helix and coil shifts. Poly L leucine has been studied by several authors[9, 36] including ourselves and most samples have shown what approximates to a single shifting peak. A very low molecular weight sample studied by the Japanese authors, however, showed a quite distinct 'double peak' behaviour.

The general conclusion regarding αCH peak shapes to be drawn from these studies of polypeptides having hydrocarbon sidechains is as follows: for real samples, i.e. those having significant polydispersity, low cooperativity in the helix–coil transition means little dependence of helicity on molecular weight and a single shifting peak is normal for all but the very lowest molecular weights. A rise in transition cooperativity means the appearance of apparent multiplicity in the peak in the helix–coil transition region, particularly for low average molecular weights. For polymers having a highly cooperative transition the 'double peak' appearance will be typical and high average

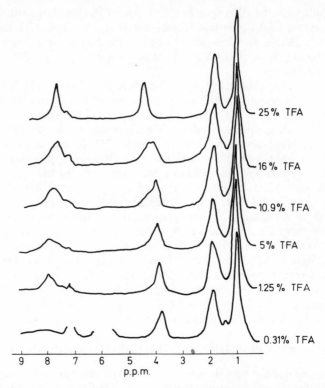

25% TFA

16% TFA

10.9% TFA

5% TFA

1.25% TFA

0.31% TFA

9   8   7   6   5   4   3   2   1
p.p.m.

Figure 17. 100 MHz spectra of poly D α amino *n* butyric acid. Sample No. 372 in TFA–CDCl₃.

molecular weights must be achieved before all molecules in a sample have the same helicity and a single shifting αC*H* peak is observed.

POLYPEPTIDES WITH AROMATIC SIDECHAINS

Poly L tyrosine (PLT)

The study of the conformation of PLT in both aqueous and non aqueous solutions by ORD and CD has been subject to uncertainty due to the presence of sidechain chromophores overlapping those of the peptide groups. In aqueous solution a marked change occurs in the ORD/CD as the pH is lowered in the range 11.5 to 11.25 and this has been attributed to both a coil → helix[37] transition and to a coil → β-structure transition[38]. In most organic solvents PLT has been assumed to be helical. We have studied the n.m.r. spectrum of PLT[16] to obtain more information on the nature of the conformational transitions and to investigate the possibility that sidechain tyrosyl–tyrosyl interactions are responsible for the overlying Cotton effects in the ordered form. The spectrum of PLT in dimethylsulphoxide (DMSOd₆) exhibits relatively sharp resonances for all protons, including the αC*H* and amide N*H* suggesting a random coil structure. Addition of water to such a solution causes a marked change in optical rotation between 5 and 10 per cent D_2O and this is accompanied by changes in the n.m.r. spectrum (*Figure 18*). The αCH resonance moves upfield by about 0.3 p.p.m. over the range of 0 to 15 per cent D_2O and thereafter remains constant in shift

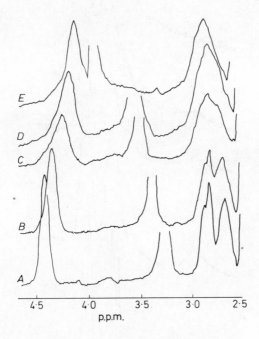

Figure 18. 220 MHz spectra of poly L tyrosine in $DMSOd_6$–D_2O mixtures at 55°C. (A) pure $DMSOd_6$; (B) 3.9% D_2O; (C) 5.7% D_2O; (D) 6.7% D_2O; (E) 18.7% D_2O.

as more water is added. Some indication of αCH multiple-peak behaviour is seen in the middle of the transition and the changes have the character expected of a coil to helix change in a polymer of low cooperativity. Some broadening of the aromatic protons also takes place over the same solvent range with little change in chemical shift. There is strong evidence that PLT is helical in trimethylphosphate (TMP)[39, 40] and addition of TMP to a solution of PLT in $DMSOd_6$ causes a sharp change in the ORD at ~ 35 per cent TMP, similar to that observed on D_2O addition. *Figure 19* shows the changes in the aromatic spectrum and as with D_2O addition there is considerable broadening with little change in chemical shift. Thus there is a strong indication that PLT is random coil in DMSO and that there is transition on D_2O or TMP addition to a more rigid conformation, probably the α-helix. However, there are no marked changes in aromatic chemical shifts as would indicate strong tyrosyl–tyrosyl interactions.

75

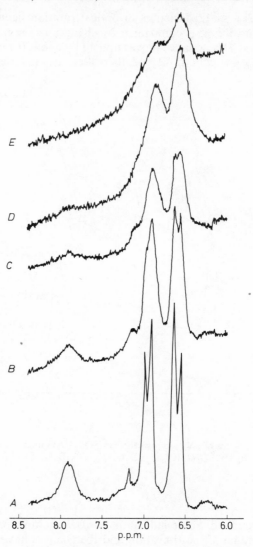

Figure 19. 100 MHz spectrum of poly L tyrosine in DMSOd₆–TMP at 29°C. %TMP: (A) 0; (B) 2.0; (C) 2.5; (D) 30; (E) 60.

Poly L aspartate esters

The conformations of ester derivatives of poly L aspartic acid have been shown to be dependent on the precise nature of the sidechain and on the solution conditions[41]. Thus whereas poly β benzyl L aspartate and poly β methyl L aspartate take up the left handed (LH) helical conformation in chloroform the β ethyl, β propyl and β phenethyl esters are in the right handed (RH) helical form. Furthermore, copolymers of a LH-supporting

76

sidechain, e.g. benzyl, with a RH-supporting sidechain, e.g. ethyl, undergo transition from the RH helical form to the LH helical form as the temperature is raised. The temperature at the midpoint of the transition is dependent on the copolymer composition. This delicate balance of helical senses and the relative ease with which either helix sense can be broken down to the coil by means of haloacetic acids indicate that the intramolecular forces differ markedly from those in such polypeptides as PBLG. Scheraga and his co-workers[42] have calculated the sidechain conformations corresponding to minimum energy of both helix senses of several poly L aspartates. They find four such conformations, two longitudinal and two transverse with respect to the helix direction. The preferred helix sense for each polymer follows from these calculations and usually is found to agree with that observed. The n.m.r. spectra of these polymers have been studied[43] to follow the variation in helix sense of L-aspartates (using the shift of the two main chain protons) and also to investigate the proposals[42] of specific sidechain conformations (using principally the spectrum of the βCH_2 group).

MAIN CHAIN RESONANCES

The effect of helix sense on the n.m.r. spectrum can be readily investigated by inclusion of L alanine residues in poly β benzyl L aspartate since as little as ten per cent L alanine is sufficient to swing the helix sense from LH to RH. *Figure 20* shows the spectrum of both helix senses in $CDCl_3$ (the 0.5 per cent

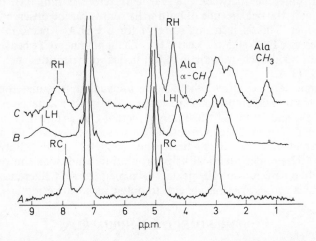

Figure 20. 100 MHz spectra of poly β benzyl L aspartate: (a) random coil (RC) form in 5% TFA–95% $CDCl_3$; (b) left handed (LH) helical form in 0.5% TFA–99.5% $CDCl_3$; (c) right handed (RH) helical form of poly [β benzyl L aspartate (90)–L alanine (10)] in 0.5% TFA–99.5% $CDCl_3$.

TFA is added to avoid aggregation broadening) and the random coil in $CDCl_3$–5 per cent TFA. The amide NH resonances is seen to be markedly dependent on conformation[56]. For PBLA the shift values from internal TMS are as follows: coil–8.00 p.p.m. (PBLG coil 7.95 p.p.m.), LH helix—8.75 p.p.m.,

77

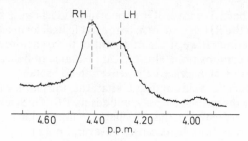

Figure 21. 300 MHz spectrum of the α*CH* resonance of helical poly [β benzyl L aspartate (95)–
L alanine (5)] in 0.5 % TFA–99.5 % CDCl₃.

RH helix—8.20 p.p.m. (PBLG helix—8.17 p.p.m.). Thus although the shifts for the coil and RH helix and thus their difference, are similar to those of PBLG, the LH helix is at markedly lower field. The shift difference between the two helix senses is thus 0.55 p.p.m. for this resonance.

The PBLA α*CH* shift is also dependent on conformation; 4.85 p.p.m. for the coil (PBLG coil 4.45 p.p.m.), LH helix—4.30 p.p.m. and RH helix—4.40 p.p.m. (PBLG helix—3.95 p.p.m.). For this main chain proton therefore, neither the RH helix nor the coil shifts are similar to those of PBLG, although the RH helix–coil shift difference $\Delta_{H/C}$ is the same for both polymers. The α*CH* shift difference between the two helix senses is thus 0.10 p.p.m. and though small, the coexistence of both helix senses can be observed in a suitable polymer. This is seen in *Figure 21* for a PBLA–5 per cent L alanine copolymer in CDCl₃–0.5 per cent TFA. The L alanine α*CH* peak is seen at 3.96 p.p.m., a value characteristic of RH helical polypeptides such as PBLG in CDCl₃–TFA.

It is important to know whether the unusual shift values for the poly L aspartate main chain protons are general to all poly L aspartates. We have observed the spectrum of helical poly β methyl L aspartate (LH) in CDCl₃–0.5 per cent TFA and the N*H* resonance was observed at 8.73 p.p.m. The α*CH* resonance has previously been reported at 4.27 p.p.m.[9]. Poly β ethyl L aspartate (RH) under identical helix-promoting conditions showed an N*H* peak at 8.16 p.p.m. and an α*CH* peak at 4.4 p.p.m. The shift differences between the two helix senses are thus general to poly L aspartate esters.

THE SIDECHAIN SPECTRUM

The sidechain spectra of poly L aspartates particularly that of PBLA in the LH helical form have been analysed[43] in terms of both shifts and coupling constants, largely with a view to establishing whether in the helical form the sidechain is at all rigidly held in preferred conformations.

Figure 20 (at 100 MHz) shows that in the coil form of PBLA the two β protons are roughly equivalent, as are the two benzyl protons. Spectra at 220 MHz reveal small shift differences (and hence essentially AB quartets) for both pairs: $\Delta \approx 0.07$ p.p.m. (βCH₂) and $\Delta \approx 0.05$ p.p.m. (benzyl CH₂). Such small differences are not unexpected, bearing in mind the presence of the

78

α asymmetric centre, and do not indicate anything unusual in the random coil conformation.

In the LH helical form of PBLA (*Figure 20*) the shift difference Δ between the two β protons rises to 0.37 p.p.m.[56] whilst in the RH helical form containing 10 per cent L alanine the difference lies between 0.5 and 0.6 p.p.m. Poly β ethyl L aspartate (RH helix) shows a βCH_2 shift difference of ~0.6 p.p.m. and a dependence of this difference on helix sense seems general. As with the helix sense dependence of the main chain protons, this sidechain shift parameter is clearly of diagnostic value for helix sense, but cannot readily be used to obtain more detailed structural information. The measurement

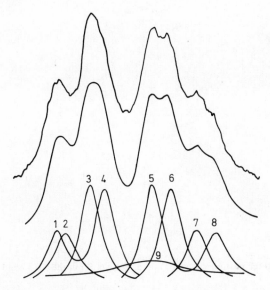

Figure 22. 100 MHz spectrum of the βCH_2 group of poly β benzyl L aspartate in $CDCl_3$ at 100°C together with a curve resolver readout and analysis.

of coupling constants, however, can in principle be made to yield detailed structural parameters. Although very difficult to observe with polypeptides in the helical form, due to the rather broad lines, we have had some success at measuring J values with PBLA at elevated temperatures. *Figure 22* shows the βCH_2 resonance at 100°C in pure $CDCl_3$[43]. It is an ABX system and the couplings to the X proton (αCH) are just visible, amounting to $J_{\alpha, A} = 4 \pm 1$ Hz, $J_{\alpha, B} = 7.4 \pm 0.3$ Hz. The three possible staggered conformations about the α—β bond can be represented as in *Figure 23* and we have attempted a conformational analysis about this bond using $J_{trans} = 13.6$ Hz and $J_{gauche} = 2.6$ Hz. Using the usual equations the three relative lifetimes are found to be $a_1 = 0.14$, $a_2 = 0.39$ and $a_3 = 0.47$, although there is the alternative solution with a_1 and a_3 interchanged, since the two β protons cannot be unequivocally assigned. Inspection of molecular models, however,

79

suggests that rotamer 2 cannot have a significant population due to steric hindrance between the sidechain carboxyl group and the helical backbone. If $a_2 = 0$, then $J_{\alpha A} + J_{\alpha B} = J_g + J_t$. Since this is not found one is led to conclude that the regular 60° rotamers of *Figure 23* are not appropriate. Inspection of the space filling model indicates that for rotamer one the $C_\alpha H_X$—$C_\beta H_B$ angle might increase by about 10° and for rotamer three the $C_\alpha H_X$—$C_\beta H_A$ angle might increase by about 15°. The application of a Karplus type function for the angular variation of the coupling constant to these

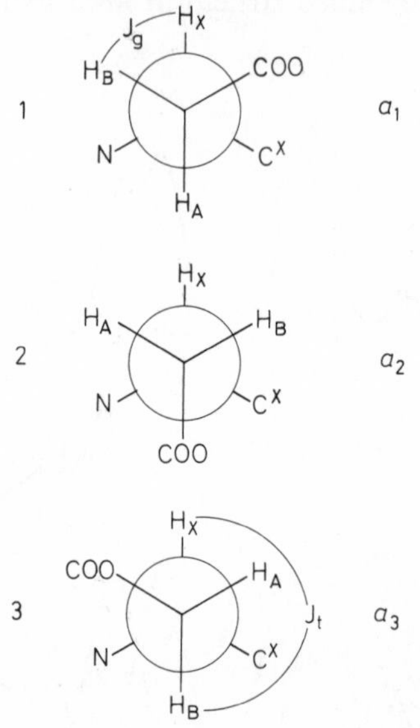

Figure 23. The three possible staggered conformations about the α—β bond of poly (aspartate esters).

'distorted' rotamers leads to an effective *trans* coupling constant of 11 to 12 Hz and an effective *gauche* coupling constant of 0.5 to 1 Hz. The sum of these quantities is not far from the sum of the observed couplings and an analysis with these constants suggests a 2:1 ratio of the 'distorted' rotamers (without of course being able to distinguish which predominates). In comparing these results with the preferred conformations of Scheraga and co-workers[42], the first conclusion that can be drawn simply by a comparison of the measured vicinal coupling constants with the postulated $\cos^2$ variation of J with α—β bond angle, is that no single rotamer can possibly satisfy the observed J values. In both the predicted Lt(−) conformation (having a transverse sidechain arrangement) and in the Ll(+) (having a longitudinal sidechain arrangement), the βCH_2 conformation approximates to that of rotamer 1. The precise bond angles indicate that one β proton should show a vicinal coupling of 12–13 Hz and the other a coupling of 1–3.5 Hz. These values are well outside the experimental error and reinforce the conclusion that under the experimental conditions the βCH_2 is in rapid motion between two well separated conformers, both having a significant lifetime. This result is not compatible with the degree of sidechain immobilization envisaged in the calculations of helix sense since the presence of a dominant sidechain conformation would certainly have been apparent from the βCH_2 spectrum.

The benzyl CH_2 spectrum of LH helical PBLA is a distinct AB quartet with $\Delta = 0.13$ p.p.m. Introduction of ten per cent L alanine induces transition to the RH form and the shift difference Δ changes to 0.28 p.p.m. The complete data on this shift difference are given in *Figure 24* together with b_0 data to indicate change of helix sense. The Δ values in both helices are much in excess of those in the coil and since the benzyl CH_2 group of PBLG under similar solvent conditions shows no measurable splitting at 220 MHz it follows that the asymmetry of the L aspartate helix is strongly felt at the benzyl

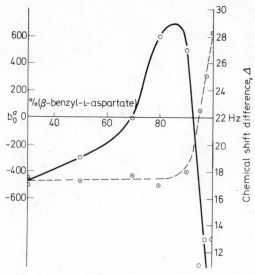

Figure 24. The shift difference Δ (at 100 MHz) between the benzyl CH_2 protons of β benzyl L aspartate residues and b_0 for the series poly [benzyl I aspartate–L alanine] in TFA–CDCl$_3$ at room temperature. Full line: shift difference; broken line: b_0.

CH_2 group. It is impossible, however, to obtain any estimate from these Δ values of the degree to which the sidechain is restricted at the βCH_2 group. Large shift differences between the benzyl protons could be generated by preferred conformations with respect to the benzene ring. However, the J_{gem} coupling of the benzyl CH_2 group in LH helical PBLA is the same as that of the random coil form and that of the monomer; this does not suggest any change in the conformation about this bond on forming the helix.

CONFORMATIONS OF ASPARTATE COPOLYMERS WITH GLUTAMATES

The αCH and amide NH chemical shifts of poly aspartate esters are both dependent on helix sense and well separated from those of poly benzyl glutamate esters. This allows separate observation of the two types of residue in block and random copolymers and thus a determination of their conformamation. Such an n.m.r. approach avoids reliance on a parameter such as b_0 that gives simply a sum of the conformations of both types of residue. Several series of A—B block copolymers (where A represents PBLA and B represents PBLG) were studied in CDCl$_3$–TFA[44]. In polymer Series 455 the L aspartate block was synthesized first, being held constant at an estimated 100 residues, and the L glutamate block of varying length was then added on without isolation of the first block. In Series 449 the order was reversed, the L gluta-

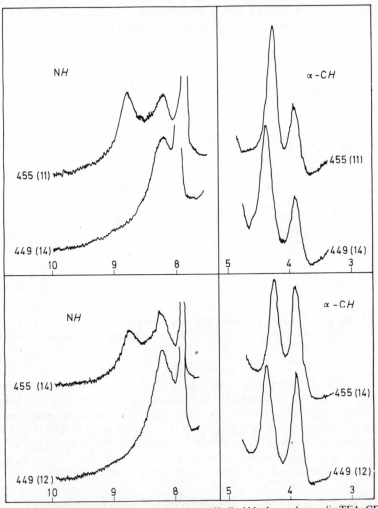

Figure 25. 220 MHz spectra of αC*H* and N*H* regions of helical block copolymers (in TFA–CDCl$_3$). Series 455 (left-handed aspartate) No. 11 (100 aspartate: 40 glutamate), No. 14 (100 aspartate: 100 glutamate); Series 449 (right-handed aspartate) No. 14 (200 aspartate: 100 glutamate) and No. 12 (100 aspartate: 100 glutamate).

mate block being synthesized first. *Figure 25* shows spectra of two fully helical polymers from both series. From the chemical shift data presented above the L aspartate αCH shift of 4.30 p.p.m. and the amide NH shift of 8.80 p.p.m. in Series 455 polymers indicates a LH helical poly β benzyl L aspartate block. This is the expected conclusion since PBLA normally takes up the LH helix. In Series 449 however, the L aspartate αCH is at 4.40 p.p.m. and the amide NH at ∼8.2 p.p.m.: this indicates a RH helical poly β benzyl L aspartate block. The most reasonable explanation for this is that there is a certain degree of overlap from the first synthesized block into the second block (as a result of some unreacted monomer still remaining). For Series 455

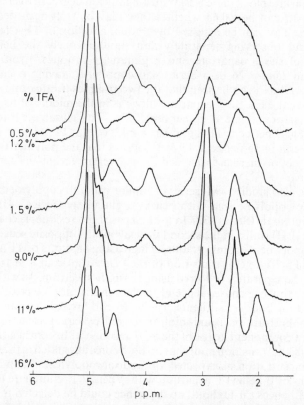

Figure 26. 100 MHz spectra, 1–6 p.p.m., of 450 (No. 14) (100 glutamate: 150 aspartate) in TFA–CDCl$_3$. Assignments: 2.1 p.p.m. glutamate βCH_2; 2.4 p.p.m. glutamate γCH_2; 3 p.p.m. aspartate βCH_2; 4–5 p.p.m. αCHs and 5.05 p.p.m. benzyl CH_2s.

this means aspartate in the L glutamate block, but an L aspartate block free of L glutamate and therefore left handed. The reverse is true for Series 449 in that the L glutamate can overlap into the L aspartate block. Since ∼15 per cent PBLG introduced into PBLA can cause a reversal of the helix sense from LH to RH[45], it is understood why the PBLA block of Series 449 is right handed.

83

The L glutamate monomer (as N-carboxyanhydride) cannot therefore have reacted to completion before the L aspartate monomer was added. A further Series (No. 450) was studied in which the L glutamate block of standard length was synthesized first and the length of the L aspartate block was increased over a wide range. When the L aspartate block was short it was found to be RH helical but as the length of the block increased the proportion of LH helix increased as judged by the αCH spectrum. This is understandable if the amount of unreacted glutamate remaining at the moment of addition of L aspartate N-carboxyanhydride were constant for all polymers since the percentage L glutamate overlapping into the L aspartate block would then be at a maximum for the shortest L aspartate block and decrease as the length of the L aspartate block increases. With a significant degree of overlap in many copolymers it can be asked whether they may be truly regarded as block copolymers. This can be assessed by gradual addition of TFA to a $CDCl_3$ solution and observing separately and simultaneously the helix to coil transition of the L aspartate and L glutamate residues. From the αCH spectrum in *Figure 26* of a Series 450 copolymer having a considerable degree of overlap it can be seen that the two residues titrate quite separately. At 1.2 per cent TFA the L aspartate block is in the middle of its transition, which is nearly complete at 1.5 per cent TFA. The L glutamate transition has not started at 9 per cent TFA and has a midpoint at 11 per cent TFA. Both blocks are coil in 16 per cent TFA. It follows that the polymer is a genuine two-block copolymer and the degree of overlap although significant, is not excessive.

We have also studied two series of similar random copolymers[46]: benzyl L aspartate copolymerized with benzyl D glutamate (Series 441) and with benzyl L glutamate (Series 432). In the first case both residues favour the same helix sense (LH) whilst in the second they favour the opposite sense. Titration of these copolymers against gradually increasing amounts of TFA in $CDCl_3$ permits the helix to coil transition of the L aspartate and L or D glutamate residues to be separately followed using the αCH spectrum. As with the block copolymers this is a consequence of the fact that a 'double peak' behaviour is observed for both constituents. The helix–coil breakdown can be characterized by the transition midpoint as judged by equal areas for the helix and the coil component peaks of the αCH resonance. In contrast to the block copolymers, the transition midpoints of both constituents fell in the same range of TFA concentration, as expected for random copolymers. In the case of the copolymers of β benzyl L aspartate with γ benzyl D glutamate (Series 441) having LH helices throughout, no difference could be detected in the transition midpoints of the two constituents and the value fell approximately linearly from 10 per cent TFA for homopoly γ benzyl D glutamate to 1.5 per cent TFA for homopoly β benzyl L aspartate. In the case of Series 432 there is a change of helix sense from LH to RH as more than 10 per cent L glutamate is added to PBLA. This is readily seen from the shift of the L aspartate αCH as explained above. The transition midpoints for Series 432 are plotted in *Figure 27* and the helix stability is seen to pass through a minimum at the point (10 per cent L glutamate) at which there is a balance between the two helix senses. Furthermore in the 50–50 copolymer there is a distinct difference between the transition midpoints of the two components.

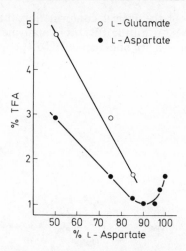

Figure 27. Helix–coil transition mid-points of Series 432: copoly [γ benzyl ʟ glutamate, β-benzyl ʟ aspartate].

This is interpreted as resulting from irregular distribution of the component residues along the chain, i.e. homopolymeric runs occur, longer than would be expected for truly random polymerization.

DL Copolymers

The conformations of racemic copolymers of random sequence have long been the subject of study, due largely to the inability of ORD/CD to provide a clear-cut answer. The intensity of the amide V band characteristic of helical polypeptides allowed[47] Tsuboi *et al.* to suggest that in the solid state PBDLG is about one half helical. The solution conformation cannot, however, be approached by this method due to the lack of suitable solvents but the α*CH* region of the n.m.r. spectrum is a direct and versatile experimental method. Bovey *et al.*[48] showed that a PBDLG sample in chloroform has an α*CH* peak at 3.95 p.p.m. (a value characteristic of helix) and moreover on TFA addition shows the 'double-peak' spectrum typical for the helix–coil transition of a low molecular weight PBLG sample. It was concluded that in chloroform the polymer was helical.

In a recent paper Paolillo *et al.*[49] have used the α*CH* spectrum of a number of DL copolymers to investigate solution conformations in several solvents. In chloroform PBDLG samples having a DP greater than about 150 were found to be fully helical, whilst the lowest molecular weight studied (DP $\approx$ 17) was less than half helical. The shift of the coil α*CH* was shown to be about 0.3 p.p.m. to lowfield of the helix. In dimethylformamide (DMF) the α*CH* showed a shift difference of 0.30 p.p.m. (helix shift 4.09 p.p.m., coil shift 4.39 p.p.m.) and whilst the PBDLG of DP = 170 (375 (1)) was fully helical, that of DP $\approx$ 17 (400A) was almost fully coil. *Figure 28* shows the results obtained for several PBDLG samples together with the corresponding spectrum of helical PBLG. In dimethylsulphoxide (DMSO) which

85

is a poorer promoter of the helical conformation than DMF (although PBLG is helical in DMSO), all the PB*DLG* copolymers were found to be in the random coil. The helicity of racemic PB*DLG* copolymers in solution was therefore shown to depend both on the solvent and on the molecular weight.

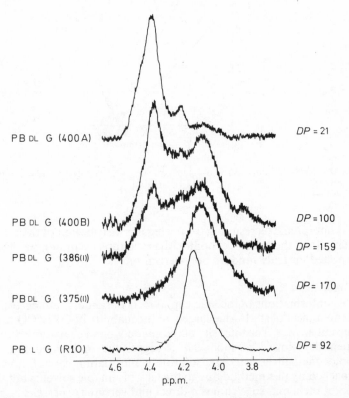

Figure 28. 220 MHz α*CH* spectra in DMF of four samples of poly γ benzyl DL glutamate having differing molecular weights and one sample of poly γ benzyl L glutamate.

The same paper also considered the conformation in chloroform of racemic copolymers of poly β benzyl aspartate (PBDLA) and of the methyl ester (PMDLA). L Aspartate residues on LH helices (which is the natural sense for both the benzyl and methyl esters) give rise to a peptide N*H* peak at 8.75 p.p.m.[43], i.e. well separated from that of the RH helix or coil conformations at 8.2–8.3 p.p.m. The N*H* region proved therefore to be of the greater diagnostic value. A PBDLA sample of DP ~60 showed an N*H* peak at 8.75 p.p.m. having less than half the total intensity of the N*H* resonance region. It was concluded that this sample is more than half in the coiled conformation. This was confirmed by the fact that on TFA addition the α*CH* resonance showed no sign of the double-peak phenomenon and moved downfield as a single line. By the same criteria as these, the sample of PMDLA (DP ~140) studied was concluded as being fully coil.

86

¹³C SPECTROSCOPY OF SYNTHETIC POLYPEPTIDES

Preliminary investigations of the helix–coil transition of PBLG have been made by FT spectroscopy at natural abundance[50]. As with many such polymer studies it was hoped that in comparison with proton spectra the lines would be very much sharper and the dependence of chemical shifts on configurational and conformational differences would be very much greater. *Figure 29* shows proton decoupled ^{13}C/FT spectra of 15 % w/v solutions of

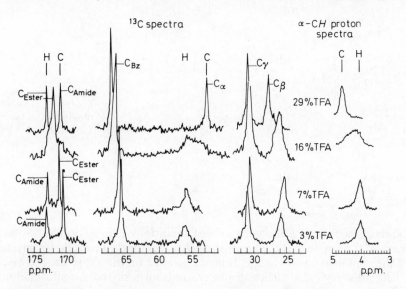

Figure 29. ^{13}C spectra at 25.2 MHz and ^{1}H spectra at 100 MHz of poly (γ benzyl L glutamate) in TFA–CDCl$_3$: 3 % and 7 % TFA—helix; 16 % TFA—helix plus coil; 29 % TFA — coil. H, helix; C, coil shifts for C$_{amide}$, C$_\alpha$ and αCH resonances.

PBLG in CDCl$_3$–TFA solvents and *Table 2* gives the observed shifts for the coil form in 29 per cent TFA and the helix form in 3 per cent TFA.

Several striking changes in chemical shift take place as the TFA concentration increases and helix is converted into coil. The α carbon moves upfield by 3.5 p.p.m. over the transition and in the intermediate region (16 per cent TFA) shows a multiple peak behaviour very similar to that seen in the proton spectrum.

It has been shown above that this is a consequence of end effects and polydispersity. The other main chain carbon atom, the amide carbonyl, likewise shows an upfield shift, of 2.5 p.p.m. in this case. The β carbon moves downfield by about 1 p.p.m. over the transition and the γ carbon remains approximately constant in shift. These displacements appear to be a direct consequence of the conformational transition since change of TFA concentration between three and seven per cent has no effect on the peak positions. This is in contrast to the benzyl carbon and the ester carbonyl resonances that show a monotonic displacement downfield with increasing

87

TFA from the very first addition. This is clearly due to solvation of the ester group by TFA, presumably by hydrogen bonding to the carbonyl group. Solvation of the ester carbonyl group by TFA results in a downfield shift of about 2 p.p.m. and if a similar value is appropriate to solvation of the

Table 2. ^{13}C Chemical shifts in p.p.m. from internal Me$_4$Si of poly (γ benzyl L glutamate) solutions in chloroform–trifluoroacetic acid

Solvent		C_β	C_γ	C_α	C_{BZ}
3% TFA 97% CDCl$_3$ (15% PBLG)	HELIX	25.9	30.7	56.3	65.9
29% TFA 71% CDCl$_3$ (15% PBLG)	COIL	27.1	30.6	53.2	67.6
Solvent		$C_{2,6-}$ Ar	C_{1-} Ar	C=O (Ester)	C=O (Amide)
3% TFA 97% CDCl$_3$ (15% PBLG)	HELIX	126.6	133.6	170.5	173.1
29% TFA 71% CDCl$_3$ (15% PBLG)	COIL	126.6 127.1	133.6	173.2	171.1

amide carbonyl under the same solvent conditions, then an upfield displacement of the amide carbonyl over the helix–coil transition could be intrinsic to the conformational change. It is not possible on the above data to assess whether the upfield α carbon displacement is intrinsic to the conformational difference since the effects of TFA solvation on this peak are not defined.

Comparison of the proton spectra of PBLG (e.g. *Figure 1*) with the ^{13}C spectra of *Figure 29* reveals an unexpected overall similarity, particularly when comparing the α carbon and α hydrogen peaks. Thus the α carbon widths (~ 65 Hz for the helix and ~ 17 Hz for the coil) are not dramatically less than peak displacements due to the conformational change from coil to helix (~ 76 Hz). Since the helix and coil are ~ 50 Hz apart in the αCH proton spectrum at the same applied field, the α carbon spectrum cannot provide substantially new information on exchange rates between the helix and coil conformations. Although the situation would be somewhat improved by working at higher fields the relatively broad lines in the carbon spectrum will make it difficult to distinguish small structural differences (as has been done for small organic molecules in solution) or to study proteins in large rigid conformations with much greater ease than is possible with the proton spectrum. The carbonyl resonances are much sharper than the α carbon and thus of potentially greater value, though often the presence of additional peaks in this region complicates the issue. Carbonyl peaks have the disadvantage of no intensity enhancement in proton decoupled spectra. Similar results to our own were subsequently reported by Boccalon *et al.*[51] on poly N δ carbobenzoxy L ornithine in CDCl$_3$–TFA, the main chain carbons showing similar displacements over the helix–coil transition as PBLG. The sidechain urethane carbonyl is markedly upfield of the amide carbonyl in this polymer and shows downfield solvation shifts of similar magnitude to those of the PBLG ester carbonyl.

We have also studied the ^{13}C spectrum of poly γ benzyl L glutamate in DMSO to observe the helical shifts and the racemic PB *DLG* to observe the coil shifts. Great similarity to the CDCl$_3$–TFA spectra was found viz. the helix αC peak was 4.5 p.p.m. to low field of the coil, the helix amide carbonyl was 4.2 p.p.m. to low field of the coil and the helix β carbon was 1.6 p.p.m. upfield of the coil.

Measurements have also been made of T_1 values for the PBLG carbon atoms under three solvent conditions in CDCl$_3$–TFA (corresponding to aggregated helix, to non-aggregated helix and to coil), using the 180°–τ–90° pulse sequence. The results, together with estimates of T_2 from linewidths are given in

Table 3. ^{13}C relaxation times in seconds, for poly γ benzyl L glutamate (30% solution in CDCl$_3$–TFA)

$$\begin{array}{c} \text{C}{=}\text{O} \qquad\qquad \text{O} \\ | \qquad\quad \beta \quad\ \ \gamma \quad\ \| \qquad \text{Bz} \\ \alpha\text{CH}{-}\text{CH}_2{-}\text{CH}_2{-}\text{C}{-}\text{O}{-}\text{CH}_2{-}\varphi \\ | \\ \text{NH} \end{array}$$

	Aggregated helix 3% TFA)		Helix 12% TFA)		Coil 50% TFA)	
	$T_1{}^*$	$T_2{}^*$	T_1	T_2	T_1	T_2
αC	~0.03	<0.003	0.1	0.005	0.06	0.02
βC	~0.03	<0.003	0.05	0.007	0.04	0.01
γC	~0.03	~0.003	0.05	0.01	0.05	0.02
BzC	0.1	~0.005	0.1	0.02	0.1	0.04
φC$_1$	3.3		3.0		3.6	
φC$_2$–C$_6$	0.8		0.6		2	
Amide C$=$O	0.9	~0.005	0.9	0.02	0.7	0.03
Ester C$=$O	1.9		1.4	0.03	1.6	0.05

*T_1 from 180–τ–90 sequence; T_2 from linewidth.

Table 3. The T_2 results are rough and moreover maximum values, since apparent linewidths will exceed true linewidths due both to the presence of structural heterogeneity and to the use of a transform of finite length. Nevertheless, the T_2 values are always less than T_1, particularly for the main chain carbons. T_1 remains roughly constant as the conformation changes due to more TFA addition, whilst T_2 values increase. This would suggest that correlation times must be of the order of a Larmor period, i.e. in the region of the T_1 minimum. Careful checking however, by measurements of T_1 at different applied fields, is required before any firm statements about correlation times can be made.

Several poly L aspartate polymers and copolymers in CDCl$_3$–TFA have been studied to investigate the dependence of the α carbon shift on helix sense and to discover whether the L aspartate shifts differ from those of the corresponding L glutamates, as found in the proton spectra. The polymers used (in addition to PBLG) were PBLA and copolymers of PBLA with PBDG (LH helical L aspartate) and with PBLG (RH helical L aspartate). The absolute magnitudes of the α carbon L aspartate shifts differ from those of the L glutamate and moreover there is a dependence on helix sense. The results are summarized in *Table 4* in p.p.m. from internal TMS.

Table 4

		αC shift			αC shift
PBLG	Helix	57.3	PBLA	LH helix	51.3
PBLG	Coil	54.3	PBLA	Coil	50.6
PBLA	RH helix	53.7			

It is seen that although the L aspartate shifts are all upfield of the L glutamate, the difference between RH L aspartate helix and coil (3.1 p.p.m.) is close to the difference between L glutamate helix and coil (3.0 p.p.m.), the coil being upfield of the helix in both cases. In both the peptide NH and αCH proton spectra the helix–coil differences $\Delta_{H/C}$ are similar for RH L aspartates and RH L glutamates, despite differences in absolute shift (see *Table 1*). The LH L aspartate helix gives a very anomalous NH shift, however. In the α carbon spectrum the LH L aspartate shift is only 1.4 p.p.m. upfield of the RH L aspartate α carbon and the LH helix–coil difference is thus reduced to 0.7 p.p.m. The LH L aspartate helix appears anomalous therefore in both the ^{1}H and ^{13}C spectra. These differences in the carbon spectrum between glutamates and aspartates (*Table 4*) have potential uses in conformational analysis.

^{13}C SPECTRA OF POLYPEPTIDES IN WATER

It is of great interest to study water-soluble polypeptides in order to establish whether there are any peak displacements that may be reliably correlated with conformation and therefore valuable in the study of protein conformation in solution. To this end we have obtained the ^{13}C spectrum of copoly (L glutamic acid[42], L lysine[28], L alanine[30]). We have previously studied the proton spectrum[5,11] and ORD of this polymer, which shows a lesser tendency to aggregation on helix formation than is shown by homopoly L glutamic acid or homopoly L lysine. Nevertheless, the α carbon resonance of the 60 to 70 per cent helical polymer at pH 2.5 had a width of some 80 Hz. As the helicity was changed from ~70 to ~10 per cent the amide carbonyl was displaced upfield by ~2.4 p.p.m. and the α carbon upfield by ~2·0 p.p.m. Both these values are approximate due to the fact that both peaks appeared multiple, perhaps due to differences between the component amino acids. In terms of a complete helix to coil change these differences would amount to about 3.5 p.p.m. for the amide carbonyl and about 3.0 p.p.m. for the α carbon. Recently the helix–coil transition of poly L glutamic acid has been studied as a function of pH in water by Lyerla et al.[52]. They were able to make measurements down to pH 4.6, at which the polymer was about three quarters helical, below which aggregation broadening precluded observations. As the pH was lowered from neutral to 4.6, both the α carbon and the amide carbonyl resonances were displaced about 2 p.p.m. downfield. In terms of a full helix–coil transition, this would represent a shift difference of about 3 p.p.m. This is close to the values obtained for the Glu–Lys–Ala copolymer.

The sum total of the results to date therefore suggests a helix–coil shift difference $\Delta_{H/C}$ for the α carbon and amide carbonyl of about 3 p.p.m. that is largely independent of solvent. This is in marked contrast to the αCH proton spectrum for which there is considerable solvent dependence of $\Delta_{H/C}$, a parameter which is found to be the lowest of all in the solvent of greatest interest, water. These preliminary carbon spectra encourage the hope that a conformational shift difference of real diagnostic value in protein studies is to be found.

ACKNOWLEDGEMENTS

The authors gratefully acknowledge the collaboration of Drs P. A. Temussi and L. Paolillo of the L.C.F.M., of the C.N.R. of Italy, Arco Felice, Naples, Italy, and the continuing support of the Science Research Council of Great Britain. Certain of the foregoing figures first appeared in articles from this laboratory in the journals listed below. We thank the editors for permission to reproduce them. Figures 1, 2, 3–Nature; Figures 14, 15, 20–24—Macromolecules; Figures 5, 6, 11, 18, 19, 28—Polymer; Figures 25, 26, 27—Biopolymers; Figure 29 and Table 2—J.C.S. Chemical Communications.

REFERENCES

[1] E. M. Bradbury, C. Crane-Robinson, H. Goldman and H. W. E. Rattle, *Nature, Lond.* **217**, 812 (1968).
[2] (a) F. A. Bovey and G. V. D. Tiers, *J. Amer. Chem. Soc.* **81**, 2870 (1959).
(b) F. A. Bovey, G. V. D. Tiers and G. Filipovich, *J. Polymer Sci.* **38**, 73 (1959).
[3] (a) G. Schwarz, *J. Molec. Biol.* **11**, 64 (1965).
(b) R. Lumry, R. Legare and W. G. Miller, *Biopolymers*, **2**, 484 (1964).
[4] J. L. Markley, D. H. Meadows and O. Jardetzky, *J. Molec. Biol.* **27**, 25 (1967).
[5] E. M. Bradbury, C. Crane-Robinson, H. Goldman and H. W. E. Rattle, *Biopolymers*, **6**, 851 (1968).
[6] F. J. Joubert, N. Lotan and H. A. Scheraga, *Biochemistry*, **9**, 2197 (1970).
[7] (a) J. H. Bradbury and G. J. Stubbs, *Nature, Lond.* **218**, 1049 (1968).
(b) J. H. Bradbury and B. E. Chapman, *J. Macromol. Sci., Chem.* **A4**, 1137 (1970).
[8] D. I. Marlborough, K. G. Orrell and H. N. Rydon, *Chem. Commun.* 518 (1965).
[9] J. A. Ferretti *Chem. Commun.* 1030 (1967).
[10] G. Schwarz and J. Seelig, *Biopolymers*, **6**, 1263 (1968).
[11] E. M. Bradbury, B. G. Carpenter, C. Crane-Robinson and H. W. E. Rattle, *Nature, Lond.* **220**, 69 (1968).
[12] F. A. Bovey, *Pure Appl. Chem.* **16**, 417 (1968).
[13] J. C. Haylock and H. N. Rydon, *Peptides 1968*, North Holland: Amsterdam (1968).
[14] F. Conti and A. M. Liquori, *J. Molec. Biol.* **33**, 953 (1968).
[15] M. Boublik, E. M. Bradbury, C. Crane-Robinson and H. W. E. Rattle, *Europ. J. Biochem.* **12**, 258 (1970).
[16] E. M. Bradbury, C. Crane-Robinson, V. Giancotti and R. M. Stephens, *Polymer*, **13**, 33 (1972).
[17] E. M. Bradbury, C. Crane-Robinson, L. Paolillo and P. Temussi, *Polymer*, (in press) (1973).
[18] E. M. Bradbury, P. D. Cary, C. Crane-Robinson, L. Paolillo, T. Tancredi and P. A. Temussi, *J. Amer. Chem. Soc.* **93**, 5916 (1971).

[19] J. H. Bradbury and M. D. Fenn, *Austral. J. Chem.* **22**, 357 (1969).
[20] J. W. O. Tam and I. M. Klotz, *J. Amer. Chem. Soc.* **93**, 1313 (1971).
[21] J. A. Ferretti, B. W. Ninham and V. A. Parsegian, *Macromolecules*, **3**, 34 (1970).
[22] J. A. Ferretti and B. N. Ninham, *Macromolecules*, **2**, 30 (1969).
[23] R. Ullman, *Biopolymers*, **9**, 471 (1970).
[24] O. Jardetzky, *Third International Conference on Magnetic Resonance in Biological Systems,* *Warrenton, Virginia* (1968).
[25] G. C. K. Roberts and O. Jardetzky, *Advanc. Protein Chem.* **24**, 447 (1970).
[26] E. M. Bradbury, C. Crane-Robinson and H. W. E. Rattle, *Polymer*, **11**, 277 (1970).
[27] K. Nagayama and A. Wada, *Biopolymers*, (in press) (1973).
[28] E. M. Bradbury, C. Crane-Robinson, L. Paolillo and P. Temussi, *J. Amer. Chem. Soc.* **95**, 1683 (1973).
[29] E. M. Bradbury and H. W. E. Rattle, *Polymer*, **9**, 201 (1968).
[30] W. E. Stewart, L. Mandelkern and R. E. Glick, *Biochemistry*, **6**, 143 (1967).
[31] J. A. Ferretti and L. Paolillo, *Biopolymers*, **7**, 155 (1969).
[32] E. M. Bradbury, P. D. Cary and C. Crane-Robinson. Unpublished observations (1972).
[33] E. M. Bradbury, P. D. Cary and C. Crane-Robinson, *Macromolecules*, **5**, 581 (1972).
[34] A. V. Purkina, A. I. Koltsov and B. Z. Volchek, *Zh. Prikl. Spektroskopii*, **15**, 288 (1971).
[35] E. M. Bradbury, J. P. Baldwin, A. Elliott, I. F. McLuckie and R. M. Stephens. Manuscript in preparation (1973).
[36] A. Warashina, T. Iio and T. Isemura, *Biopolymers*, **9**, 1445 (1970).
[37] G. D. Fasman, E. Bodenheimer and C. Lindblow, *Biochemistry*, **3**, 1665 (1964).
[38] E. Patrone, G. Conio and S. Brighetti, *Biopolymers*, **9**, 897 (1970).
[39] V. N. Damle, *Biopolymers*, **9**, 937 (1970).
[40] F. Quadrifoglio, A. Ius and V. Crescenzi, *Makromol. Chem.* **136**, 241 (1970).
[41] E. M. Bradbury, B. G. Carpenter and H. Goldman, *Biopolymers*, **6**, 837 (1968).
[42] J. F. Yan, G. Vanderkooi and H. A. Scheraga, *J. Chem. Phys.* **49**, 2713 (1968).
[43] E. M. Bradbury, B. G. Carpenter, C. Crane-Robinson and H. Goldman, *Macromolecules*, **4**, 557 (1971).
[44] L. Paolillo, P. Temussi, E. Trivellone, E. M. Bradbury and C. Crane-Robinson, *Biopolymers*, **10**, 2555 (1971).
[45] E. M. Bradbury, A. R. Downie, A. Elliott and W. E. Hanby, *Proc. Roy. Soc. A*, **259**, 110 (1960).
[46] L. Paolillo, P. Temussi, E. M. Bradbury and C. Crane-Robinson, *Biopolymers*, **11**, 2043 (1972).
[47] M. Tsuboi, Y. Mitsui, A. Wada, T. Miyazawa and N. Nagashima, *Biopolymers*, **1**, 297 (1963).
[48] F. A. Bovey, J. J. Ryan, G. Spach and F. Heitz, *Macromolecules*, **4**, 433 (1971).
[49] L. Paolillo, P. Temussi, E. M. Bradbury and C. Crane-Robinson. Submitted to *Biopolymers*.
[50] L. Paolillo, T. Tancredi, P. Temussi, E. Trivellone, E. M. Bradbury and C. Crane-Robinson, *Chem. Commun.* 335 (1972).
[51] G. Boccalon, A. S. Verdini and G. Giacometti, *J. Amer. Chem. Soc.* **94**, 3639 (1972).
[52] J. R. Lyerla, B. H. Barber and M. H. Freedman, *Canad. J. Biochem.* (in press) (1973).
[53] C. Crane-Robinson and P. Temussi. Unpublished observations (1972).
[54] This laboratory; unpublished observations (1972).
[55] J. A. Ferretti, *Polym. Prepr. Amer. Chem. Soc., Div. Polym. Chem.* **10**, 29 (1969).
[56] H. N. Rydon, *Polymer Preprints*, **10**, 25 (1969).

NOTE ADDED IN PROOF

A recent paper on poly L alanine by M. Goodman, F. Toda and N. Ueyama [*Proc. Nat. Acad. Sci., Wash.* **70**, 331 (1973)], published since this lecture was given, gives an explanation of the αCH 'double peaks' that accords exactly with our views: viz. the main highfield peak is a helix and coil time average, whilst the small lowfield peak is pure coil. As explained in the text, we do **not** consider this a general explanation of αCH double peaks, but one peculiar to poly L alanine.

LOCAL AND OVERALL VIBRATIONS OF POLYMER CHAINS

T. Shimanouchi

*Department of Chemistry, Faculty of Science,
University of Tokyo, Japan*

ABSTRACT

Internal motions of polymer chains are classified into the local and the overall vibrations. The frequencies and modes of the latter are sensitive to the conformation of the chains. The discrimination between these two kinds of modes is explained, taking the 1,2-dichloroethane molecule as an example. The overall vibrations of n-$C_{36}H_{74}$ chain molecule, $(CH_2)_{34}$ ring molecule and polypeptide chain are discussed.

INTRODUCTION

Translational and rotational motions of biopolymer molecules have been studied in detail by means of electrophoresis, sedimentation, diffusion, viscosity, osmotic pressure and other measurements and the information on the molecular weight, molecular shape and hydration has been acquired. However, the internal motions of biopolymer molecules remain to be explored, although they may be closely related with the properties and biological functions[1, 2].

The internal motions of small molecules have been widely studied by infra-red and Raman spectroscopy. The spectra obtained provide almost full information of the vibrational states of small molecules. However, for the large molecules they give only limited information, the important parts of internal motions being left undetermined.

This fact is due to the number of degrees of freedom of the internal motion. For instance, the myoglobin molecule has 2 600 atoms and the internal freedom is 7 800. It is impossible to know full details of all these motions. In order to overcome this situation, the author suggested separating the motions into local vibrations and overall vibrations, emphasized the need for study of the latter, and calculated the frequencies of the longitudinal acoustical vibrations of the α-helix[2-4]. Peticolas and his co-workers also attempted to calculate the frequencies of these modes[1, 5].

The frequencies of the overall vibrations are highly sensitive to the length or overall shape of a large molecule and give helpful information as to the conformations of polymer chains. In the present paper the discrimination between local and overall vibrations is first described, 1,2-dichloroethane being taken as an example; then overall vibrations are discussed for paraffin and polypeptide molecules.

93

1,2-DICHLOROETHANE

The infra-red and Raman spectra of 1,2-dichloroethane have been studied in detail by Mizushima and his co-workers[6-8]. The spectra in the solid state show that the molecule takes the *trans* form. The assignments of the bands are given as shown in column 4 of *Table 1*[9]. These assignments are based on the PED elements in normal coordinate treatments. However, the mode is differently described when we calculate the atomic displacements in each vibration. The motion of the ith atom for the ath normal vibration is given by

$$\left.\begin{array}{l} x_i^a = x_i^0 + d(L_x)_{xi}^a \sin 2\pi c v_a t \\ y_i^a = y_i^0 + d(L_x)_{yi}^a \sin 2\pi c v_a t \\ z_i^a = z_i^0 + d(L_x)_{zi}^a \sin 2\pi c v_a t \end{array}\right\} \tag{1}$$

where x_i^0, y_i^0 and z_i^0 are the Cartesian coordinates of the ith atom in the equilibrium position, v_a is the wavenumber of the ath vibration, $(L_x)_{xi}^a$, $(L_x)_{yi}^a$ and $(L_x)_{zi}^a$ are the elements of the L_x matrix defined by

$$\mathbf{X} = L_x \mathbf{Q} \tag{2}$$

($\mathbf{X}$ is the Cartesian displacement vector and $\mathbf{Q}$ the normal coordinate vector.) The coefficient d is given by

$$d = 6.84 \, (T^{\frac{1}{2}}/v_a) \tag{3}$$

(T is the absolute temperature), when the L_x matrix elements are calculated using the atomic weight unit for the mass and the ångström unit for the length and when the sum of the kinetic and potential energies for each vibration is assumed to be kT. *Figures 1, 2* and *3* show the values of x_i^a, y_i^a and z_i^a, when we take $-90°$, $-80°, \ldots,$ $80°$ and $90°$ for $2\pi c v_a t$. For almost all the atomic displacements the lines drawn are overlapped and just give the area of displacements.

The result of the interpretation of these figures is given in column 5 of *Table 1*. The v_6 (CCCl symmetrical deformation) vibration is assigned to the molecule elongation mode. This is a kind of Cl . . . Cl stretching mode, other atoms following the movement of the Cl atoms. The v_{10} (CC torsion) and v_{18} (CCCl antisymmetrical deformation) vibrations are assigned to the molecule deformation modes, since the two Cl atoms move in one direction and the central CH_2—CH_2 group moves in the opposite direction. The v_{17} (CCl antisymmetrical stretching) vibration is similar. However, this mode does not cause the deformation of the whole molecule and is assigned to the local translation of the CH_2—CH_2 group.

Other modes are almost pure local vibrations. For $v_1, v_2, v_3, v_7, v_8, v_{11}, v_{12}, v_{14}, v_{15}, v_{16}$ the Cl and C atoms are almost fixed and they are just the hydrogen vibrations. For v_9 and v_{13} (CH_2 rocking) vibrations the two carbon atoms move slightly and they are assigned to the twisting and rotation modes of the CH_2—CH_2 group, respectively. For the v_4 (CC stretching) vibration the two Cl atoms are almost fixed and it is just the local CH_2—CH_2 stretching mode.

For the v_5 (CCl symmetrical stretching) vibration the two Cl atoms are

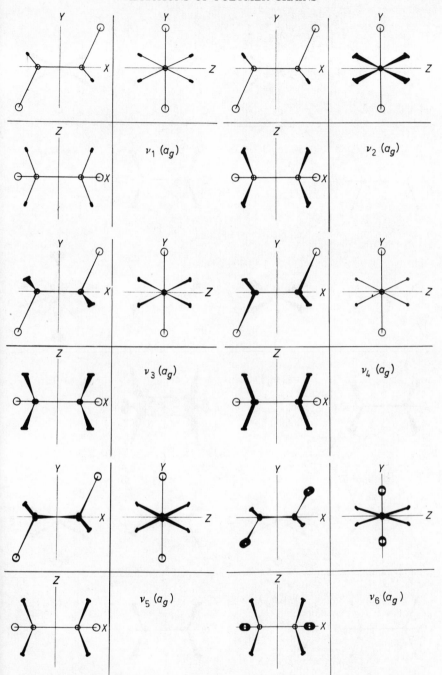

Figure 1. Atomic displacements in the normal vibrations of the 1,2-dichloroethane molecule. (The *trans* form. Classical dynamics and the harmonic approximation are used. The amplitudes are 2.5 times larger than those at 300 K.)

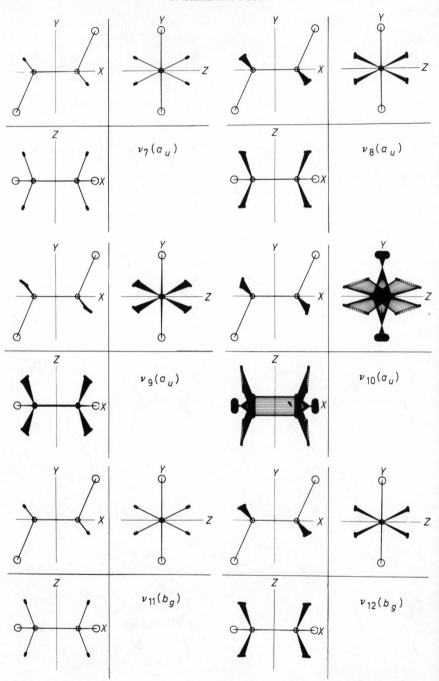

Figure 2. Atomic displacements in the normal vibrations of 1,2-dichloroethane (continued).

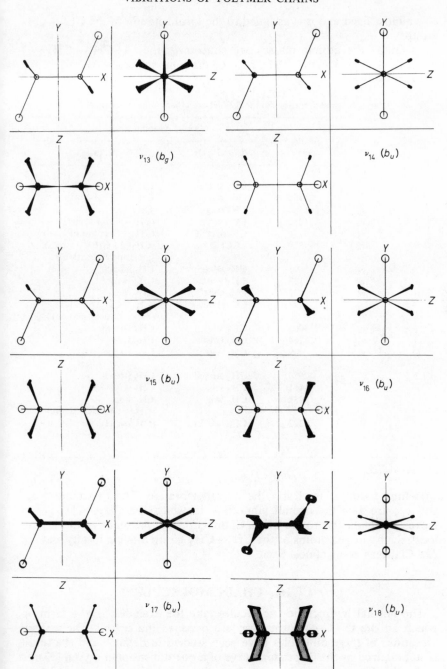

Figure 3. Atomic displacements in the normal vibrations of 1,2-dichloroethane (continued).

97

also almost fixed and it is assigned to the local rotation of the CH_2—CH_2 group.

As far as the atomic masses are concerned, the Cl—CH_2—CH_2—Cl molecule is approximately the $Cl \ldots Cl$ diatomic molecule. The $Cl \ldots Cl$

Table 1. Normal vibrations of the 1,2-dichloroethane molecule (transform)

| | | obs. cm^{-1} | Assignments | |
			(from PED)	(from L_x)
a_g	ν_1	2957	CH_2 stretch.	CH_2 stretch.
	ν_2	1445	CH_2 sciss.	CH_2 sciss.
	ν_3	1304	CH_2 wag.	CH_2 wag.
	ν_4	1052	CC stretch.	(CH_2—CH_2) stretch.
	ν_5	754	CCl stretch.	(CH_2—CH_2) rotation.
	ν_6	300	CCCl deform.	(CH_2Cl—CH_2Cl) stretch. (molecule elongation)*
a_u	ν_7	3005	CH_2 stretch.	CH_2 stretch.
	ν_8	1123	CH_2 twist.	CH_2 twist.
	ν_9	773	CH_2 rock.	CH_2 rock.
	ν_{10}	123	CC torsion	(CH_2—CH_2) translation (molecule deformation)*
b_g	ν_{11}	3005	CH_2 stretch.	CH_2 stretch.
	ν_{12}	1264	CH_2 twist.	CH_2 twist.
	ν_{13}	989	CH_2 rock.	CH_2 rock.
b_u	ν_{14}	2983	CH_2 stretch.	CH_2 stretch.
	ν_{15}	1461	CH_2 sciss.	CH_2 sciss.
	ν_{16}	1232	CH_2 wag.	CH_2 wag.
	ν_{17}	728	CCl stretch.	(CH_2—CH_2) translation
	ν_{18}	222	CCCl deform.	(CH_2—CH_2) translation (molecule deformation)*

* Overall vibrations.

stretching mode ν_6 is clearly the overall vibration. The two modes, ν_{10} and ν_{18} are also the overall vibrations, in which the CH_2—CH_2 group moves perpendicular to the $Cl \ldots Cl$ direction. All the other vibrations are local, the hydrogen atoms or the CH_2—CH_2 group moving locally and the two Cl atoms being almost fixed.

n-$C_{36}H_{74}$ CHAIN MOLECULE

The normal hydrocarbon molecules take the extended zigzag form, in which all the C—C conformations are *trans*, in the crystal. The normal vibrations of these molecules have been studied in detail[10] and the result is summarized as the dispersion curve of a one-dimensional crystal (*Figure 4*). As this figure shows, the ν_5 and ν_9 vibrations are acoustical modes, to which the overall vibrations belong. All the other vibrations appearing in the region higher than 700 cm^{-1} are optical modes, which correspond to the local vibrations.

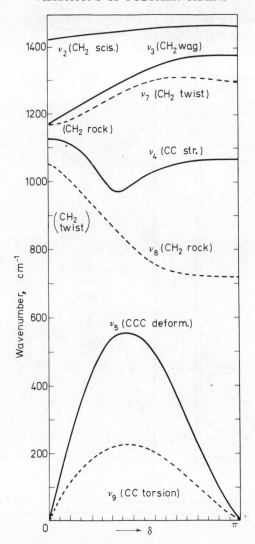

Figure 4. Frequency versus phase difference (δ) relationship for polymethylene chain (v_4 and v_5 are coupled strongly and v_5 vibrations with small phase differences are assigned to the CC stretching mode. See text. The curves for v_1 and v_6 appearing in the 3000 cm^{-1} region are omitted.)

Figure 5 shows the Raman spectra of n-$C_{36}H_{74}$ and the correspondence between the spectra and the dispersion curve. A progression of bands appearing in the low frequency region is the longitudinal acoustical vibrations[11]. The lowest frequency band at 67.4 cm^{-1} is assigned to the accordion-like vibration.

99

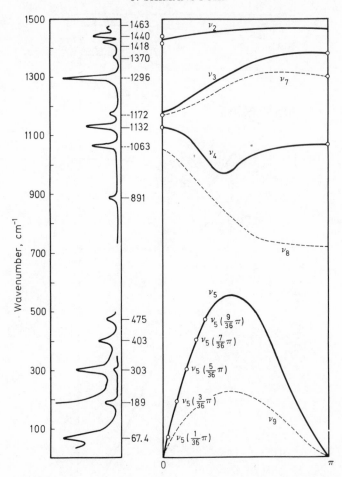

Figure 5. Raman spectra and band assignments of n-$C_{36}H_{74}$ in the crystalline state. (The 891 cm^{-1} Raman band is assigned to the methyl group rocking vibration.)

This vibration was first found and assigned by Mizushima and the author[12] in 1944 for the smaller normal paraffin molecules and the frequencies are expressed by

$$v = 2\,400/N_C \text{ cm}^{-1} \qquad (4)$$

where N_C is the number of carbon atoms. The form of the equation is based on

$$v = (1/2l)(E/\rho)^{\frac{1}{2}} \qquad (5)$$

where v is the frequency of the accordion-like vibration of a rod, the length, the density and Young's modulus of which are l, ρ and E, respectively.

Equation 4 gives the frequency accurately, when the number of carbon atoms is large. In *Table 2* the frequencies calculated from equation 4 are compared with those observed by Schaufele and the author in 1967[11]. This equation is useful for the study of the length of a hydrocarbon chain[11] and also the thickness of the single polyethylene crystal[13].

Table 2. Accordion-like vibrational frequencies of polymethylene chains utilizing equation 4

No. of C atoms	Calc. (cm^{-1})	Obs.* (cm^{-1})	No. of C atoms	Calc. (cm^{-1})	Obs.* (cm^{-1})
18	131	133	32	75	76
20	120	114	36	67	67
24	100	98	44	55	57
28	86	85	94	26	26

* See ref. 11.

The accordion-like vibration is a typical overall vibration. Other parts of the dispersion curve for v_5 and v_9 in *Figure 4* are related with other kinds of overall vibrations. The modes of these vibrations are given in *Figure 6*, those

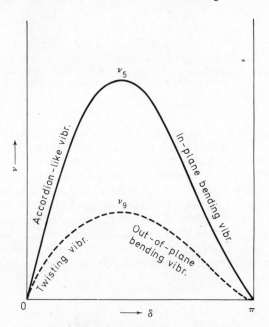

Figure 6. Frequency versus phase difference curves for acoustical vibrations of the polymethylene chain and their vibrational modes as overall vibrations.

of the lowest frequency ones are drawn in *Figure 7*, and their frequencies are given in *Figure 8* as a function of the number of carbon atoms.

In *Figures* 9 and *10* a couple of the overall vibrations of n-$C_{36}H_{74}$ are shown as examples. The modes given in *Figure 5* and the PED elements[14] show that the accordion-like vibration belongs to the CC stretching mode and not to the CCC bending mode.

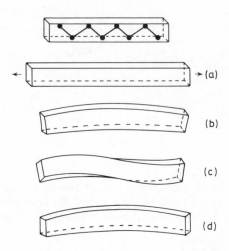

Figure 7. Lowest frequency modes for four kinds of overall vibrations of the polymethylene chain cited in *Figure 6.* (a) accordion-like vibrations, (b) in-plane bending vibrations, (c) twisting vibration and (d) out-of-plane bending vibration.

The phase differences of the one-dimensional crystal are the problem of interest and are discussed by Shimanouchi and Tasumi[14] and by Peticolas *et al.*[5]. When the extended or helix chain consists of n units, the phase difference found for neighbouring units is given by $m\pi/(n + 1)$, where m is the order of vibration and $m = 1$ for the accordion-like vibration. Accordingly, the phase difference for the CC stretching mode is given by $m\pi/n_C$, since the chain has $n_C - 1$ CC bonds. Similarly the phase differences for the CCC bending mode and for the CC torsional mode are given by $m\pi/(n_C - 1)$ and $m\pi/(n_C - 2)$, respectively. In accord with the above assignment the phase differences for the normal paraffin molecules are given by $m\pi/n_C$ better than by $m\pi/(n_C - 1)$[14].

$(CH_2)_{34}$ RING MOLECULE

The laser-Raman scattering by $(CH_2)_{34}$ was measured and the normal vibrations were calculated by Schaufele, Tasumi and Shimanouchi[15, 16]. This molecule takes approximately the form given in *Figure 11* and has overall vibrations more complex than those of the n-paraffin chains. *Figure 12* gives some of them which correspond to the observed low frequency Raman bands. The accordion-like vibrations of this rectangular molecule are expected to have frequencies close to that of $n\text{-}C_{15}H_{32}$. The frequency calculated for the latter is 160 cm^{-1} which is not far from the observed value, 145 cm^{-1}.

POLYPEPTIDE CHAIN

The overall vibrations of protein molecules may be very complex and are one of the aims of this series of researches. So far no observed Raman band has been assigned to this sort of vibration. The only way of estimating the modes and the frequencies is to transfer the force constants from small molecules with similar structures and to calculate the normal modes and the frequencies. Calculations have been made for the right-handed α-helix of poly-L-alanine and the relation between the length of the helix and the frequency of the accordion-like vibration was estimated[3, 5]. The helix rod

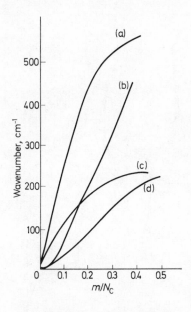

Figure 8. Frequencies of four kinds of overall vibrations of the polymethylene chain as functions of the order of vibration m and the number of carbon atoms n_C. As for (a), (b), (c) and (d), see *Figure 7.*

has other overall vibrations similar to those shown in *Figure 7*. The frequencies of these vibrations are also functions of the length of the rod and are estimated from the low frequency dispersion curve[3, 5]. Some of the overall vibrations of protein molecules like myoglobin or lysozyme may be similar to those shown in *Figure 12*. However, there may be other vibrations the amplitudes of which are considerably **larger.**

Figure 9

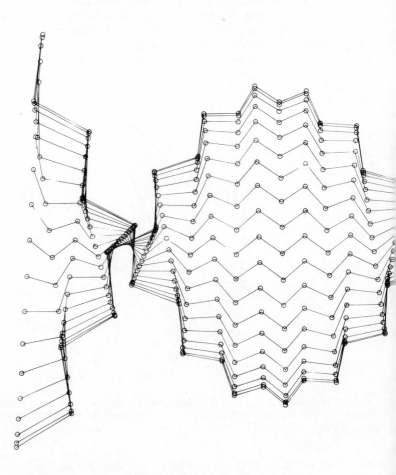

Figure 10

Figure 9. Atomic displacements of $C_{36}H_{74}$ for the accordion-like vibration at 300 K (Classical dynamics and the harmonic approximation are adopted and equation 1 is used. The methylene groups are treated as one dynamical unit.)

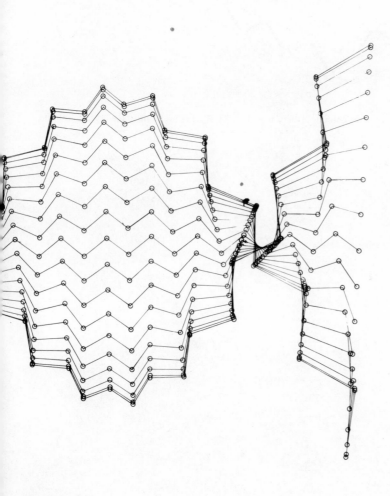

Figure 10. Atomic displacements of $C_{36}H_{74}$ for the lowest frequency 'gerade' mode belonging to the in-plane bending vibration at 300 K. (See *Figure 9*. The calculated frequency is 12 cm^{-1}.)

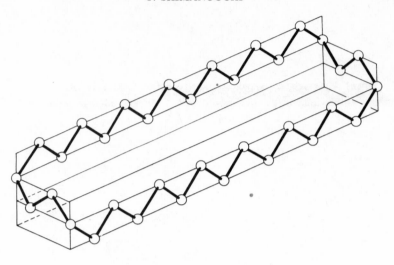

Figure 11. Approximate structure of the $(CH_2)_{34}$ ring molecule in the crystalline state.

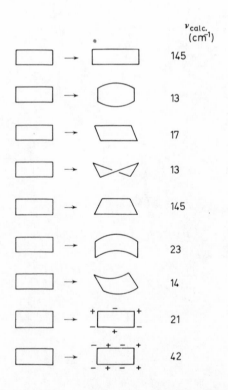

Figure 12. Overall vibrations of the $(CH_2)_{34}$ molecule.

REFERENCES

[1] W. L. Peticolas and M. W. Dowley, *Nature, London*, **212**, 400 (1966).
[2] T. Shimanouchi, *Shizen 1971* No. 6, page 47.
[3] K. Itoh and T. Shimanouchi, *Biopolymers*, **9**, 383 (1970).
[4] T. Shimanouchi, *Disc. Faraday Soc.* **49**, 60 (1970).
[5] B. Fanconi, E. W. Small and W. L. Peticolas, *Biopolymers*, **10**, 1277 (1971).
[6] S. Mizushima, *Structure of Molecules and Internal Rotation*. Academic Press: New York (1954).
[7] *Scientific Papers of Professor S. Mizushima*, University of Tokyo (1959).
[8] S. Mizushima, T. Shimanouchi and I. Harada, *Commentarii* (Pontifical Academy of Sciences), Vol. II, No. 43.
[9] T. Shimanouchi, *Tables of Molecular Vibrational Frequencies*, consolidated volume, NSRDS-NBS 39 (1972).
[10] For example, M. Tasumi, T. Shimanouchi and T. Miyazawa, *J. Molec. Spectrosc.* **9**, 201 (1962); **11**, 422 (1963).
[11] R. J. Schaufele and T. Shimanouchi, *J. Chem. Phys.* **47**, 3605 (1967).
[12] S. Mizushima and T. Shimanouchi, *Proc. Imp. Acad. Tokyo*, **20**, 86 (1944); *J. Amer. Chem. Soc.* **71**, 1320 (1949).
[13] W. L. Peticolas, G. W. Hiblar, J. L. Lippert, A. Peterlin and H. Olf, *Appl. Phys. Letters*, **18**, 87 (1971).
[14] T. Shimanouchi and M. Tasumi, *Indian J. Pure Appl. Phys.* **9**, 958 (1971).
[15] M. Tasumi, T. Shimanouchi and R. F. Schaufele, *Polymer J.* **2**, 740 (1971).
[16] R. F. Schaufele and M. Tasumi, *Polymer J.* **2**, 815 (1971).

BARRIERS AND CONFORMATIONS

William G. Fateley

*Department of Chemistry, Kansas State University,
Manhattan, Kansas 66506, USA*

ABSTRACT

The theoretically calculated barriers for the OH rotor in parasubstituted phenols are in very good agreement with those found experimentally. Further measurements on a large number of monosubstituted phenols provide sufficient data to allow the prediction of barriers and torsional frequencies in multi-substituted phenols.

Obviously, mesomeric and inductive are linear effects in phenols and this encourages the prediction of other barriers and torsional frequencies in aromatic systems. A new parameter $\Delta\omega_t$ is introduced to describe the electronic climate about the C—O bond better.

Asymmetrical potential functions are investigated for C—C, C—O, C—N and C—S single bonds. The influence of different groups, geometric changes during rotation and higher order terms are discussed. Empirical data are given for the effects of different groups on internal rotation.

PREFACE

Certainly this Conference provides a beautiful environment for asking some questions important to conformations in macromolecules:

1. Can barriers to internal rotation be predicted from:
   (a) experimentally determined parameters?
   (b) theoretical calculation?
2. Can barrier information be successfully applied to macromolecules?

Let us examine these questions in parts. First, the theoretically predicted barriers agree well with experimental results. The experimental results for barriers are found to be useful in predicting C—O barriers in phenolic systems, thus demonstrating the transferability of these data. Second, the application of barrier values to large molecules will be discussed in some detail. Interaction, geometric changes and potential surfaces will play important roles in predicting conformations.

109

WILLIAM G. FATELEY

Part I

Characterization of the Barrier about the Aromatic Carbon–Oxygen Bond in Phenols

(In collaboration with Dr G. L. Carlson)

I. INTRODUCTION†

Both theory[1] and experiment[2] predict that phenol (C_6H_5OH) is planar. A planar structure is stabilized by delocalization of the p-type lone pair on oxygen, see *Figure 1*(a); this is likely to be more effective than the delocalization of the more tightly bound sp^2-type lone pair in the orthogonal form, *Figure 1*(b). The energy difference represents the barrier to rotation about the C—O bond. This barrier is considerably larger than in aliphatic alcohols reflecting, in part, the increased double bond character in the C—O bond.

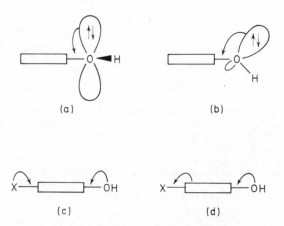

Figure 1. An end-on view of the benzene ring (▭). Part (a) shows the distribution of the p-type lone pair electrons into the bond. Part (b) illustrates the angular dislocation effect. Parts (c) and (d) demonstrate either opposition to, or reinforcement of, delocalization of the oxygen lone pair electrons by the substituent X.

Campagnaro, Hall and Wood[3] have examined far infra-red spectra of several para-substituted phenols and observed changes in the OH torsional frequency which correlated with qualitative electronic properties of the substituent†. To confirm and extend these ideas, we have carried out *ab initio* molecular orbital calculations and far infra-red spectroscopic measurements of the torsional barriers in the para-substituted phenols XC_6H_4OH (X = OH, F, CH_3, H, CHO, CN and NO_2)‡.

---

† Other spectroscopic studies of the effects of *para*-substituents have been reported for benzaldehydes[3] and anilines [D. G. Lister and J. K. Tyler, *Chem. Commun.* 152 (1966); A. Hastie, D. G. Lister, R. L. McNeil and J. K. Tyler. *Chem. Commun.* 108 (1970)].

‡ L. Radom, W. J. Hehre, J. A. Pople, G. L. Carlson and W. G. Fateley, *Chem. Commun.* 308 (1972).

Far infra-red spectra of the phenols in cyclohexane solution ($\sim 0.01\text{M}$) were obtained by procedures described elsewhere (see ref. 4 and Section II-A). The far infra-red spectra of the analogous phenol–OD compounds were also obtained to confirm the assignments of the torsional frequencies[4]. The theoretical method used is self consistent field molecular orbital theory in the linear combination of atomic orbitals (LCAO) approximation[5] using the STO-3G basis set[6] and standard geometries[7]. Orbital and overlap electron populations are calculated using Mulliken's method[8].

Results are quoted in *Table 1* where the quantities listed have the following meanings. ΔV_2 is the change† in the twofold barrier from the value in phenol, $q_\pi(X)$ is the π charge on the substituent X (a positive value indicates that X is a π donor), $\Delta q_\pi(\text{OH})$ is the change in π charge on OH relative to the value in phenol, and $\Delta \pi_{C-O}$ is the corresponding change in the double bond character of the C—O bond as measured by the π-overlap population. The interaction energy of the substituent X with the OH group is measured by the energy

$$X-\!\!\bigcirc\!\!-OH \;+\; \bigcirc \;\longrightarrow\; X-\!\!\bigcirc \;+\; \bigcirc\!\!-OH \quad (1)$$

change in the formal reaction (1). If X and OH do not interact, the energy change in (1) would be zero. Interaction energies are listed for both planar and orthogonal orientations of the OH group in (1).

There are several interesting points to note from the results in *Table 1*.

(1) Although the calculated barrier in phenol (5.16 kcal mol^{-1})[1] is considerably higher than the experimental value (3.56 kcal mol^{-1}), the changes in barrier with substitution as shown in *Table 1* are in quite good agreement.

(2) The barrier is found to decrease when X is a π-electron donor [positive $q_\pi(X)$ charges] and conversely.

(3) These results are easily rationalized in terms of the opposition, *Figure 1*(c), to, or reinforcement, *Figure 1*(d), of, delocalization of the oxygen lone pair electrons by the substituent X. Thus, when X is a π-donor as in *Figure 1*(c), electron donation by OH decreases [$\Delta q_\pi(\text{OH})$ is negative], the double bond character in the C—O bond decreases [$\Delta \pi_{C-O}$ is negative) and the barrier decreases. Conversely, when X is a π-acceptor as shown in *Figure 1*(d), electron donation by OH increases, the double bond character in the C—O bond increases and the barrier increases. The theory and experimental evidence thus confirm the conclusions reached by Campagnaro, Hall and Wood[3].

(4) The calculated values of the interaction energies show that when X is a π-donor, the interaction between X and OH is destabilizing (negative interaction energies) and conversely. This is true in both planar and orthogonal conformations although smaller in the latter. There is an almost linear correlation between the barrier values and the interaction energies in the planar forms.

---

† In all cases, we take the *change* in a calculated or experimental quantity to be the value in the substituted phenol less the value in phenol.

111

Table 1. Calculated and experimental quantities for para-substituted phenols (X—C$_6$H$_4$—OH)

X	ΔV_2 (kcal mol^{-1})[a]		$q_\pi(X)$[a,b]	$\Delta q_\pi(OH)$[a,b]	$\Delta\pi_{C-O}$[a,b]	Interaction energy[a] (kcal mol^{-1})	
	Exp.	Calc.				Planar	Orthogonal
OH	−0.87	−0.95[b]	+0.096	−0.006	−0.006	−1.4	−0.5
F	−0.60	−0.53	+0.074	−0.002	−0.003	−0.9	−0.4
CH$_3$	−0.32	−0.28	+0.008	−0.002	−0.002	−0.4	−0.1
H	0	0	0	0	0	0	0
CHO	+0.87	+0.47[b]	−0.041	+0.006	+0.004	+0.7	+0.2
CN	+0.70	+0.66	+0.029	+0.009	+0.006	+0.8	+0.2
NO$_2$	+0.98	+1.02	−0.039	+0.013	+0.009	+1.3	+0.3

[a] Defined in text. [b] Values quoted for planar conformations. Even in correlation using Hammett σ parameter, the CHO group shows poor correlation. [c] Twofold (V_2) component of barrier calculated via a two-term Fourier expansion: $V(\phi) = V_1(1 - \cos \phi)/2 + V_2(1 - \cos 2\phi)/2$. The most stable conformation when X = OH has HO…OH trans and the cis–trans energy difference is 0.07 kcal mol^{-1}. The most stable conformation when X = CHO has HO…CO cis and the cis–trans energy difference is 0.08 kcal mol^{-1}.

II. DISCUSSION

A. The case for cyclohexane solution data

It is well known that phenols are strongly associated in the liquid state and also undergo hydrogen bonding with other molecules and many solvents. With cyclohexane, there is no possibility of hydrogen bonding with the solvent†, but solute–solvent interactions have been reported. V. von Keusler[9] has shown that phenol in cyclohexane at a concentration of approximately 3×10^{-4} mol/l. is completely monomeric but at a concentration of 0.15 mol/l. it is 52 per cent associated and 48 per cent monomer. Dearden and Forbes[16] found an approximately 5 mμ shift between the u.v. spectra of the vapour and 2.3×10^{-4} mol/l. cyclohexane solutions which they attributed to solute–solvent interaction because phenol should be 100 per cent monomer at this concentration. Furthermore, Evans[10] has noted the dependence of some infra-red band positions for phenol on the degree of association.

We have ruled out any large effect of association on the torsional frequency of phenol in cyclohexane solution for two reasons. First, we have compared the torsional frequencies of several halogen-substituted phenols obtained for cyclohexane solution with the corresponding vapour phase values of Fateley, Miller and Witkowski[11]. These data are given in *Table 2*. The agreement between solution and vapour values is almost perfect except for

Table 2. Torsional frequencies of halo-substituted phenols. Comparison of vapour and cyclohexane solution values

Compound	Vapour (cm^{-1})	Torsional frequency	
		Cyclohexane solution	
		cm^{-1}	Concentration (mol/l.)
Phenol	310	310	0.010
p-Fluorophenol	280	280	0.009
p-Chlorophenol	302	302	0.015
m-Fluorophenol	311 } 319 ╎	329 } 316 ╎ *	0.009
m-Chlorophenol	313	312	0.008
m-Bromophenol	314	312	0.006

* Doubling due to the presence of *cis* and *trans* forms. These rotamers had previously been reported by us for metafluorobenzaldehyde.

meta-fluorophenol which is also peculiar in that the torsion is observed as a doublet in both solution and gas. Secondly, we have carried out a dilution study on phenol in cyclohexane solution to determine if there is any change in torsional frequency as a function of concentration. Over the concentration range 0.02 to 0.001 mol/l., which is the lower limit of detection with a 1 cm path length, the torsional frequency did not shift by more than 1 cm^{-1}.

† An exception to this observation may be the ortho-fluorophenol which might be intermolecularly hydrogen bonded.

These two observations are taken as strong indication that for phenol in cyclohexane at concentrations below 0.02 mol/l., association effects on the torsional frequency are negligible. Furthermore, the good agreement of solution and vapour data also seems to rule out any appreciable effect of solvent–solute interaction.

Similar agreement between solution and vapour phase torsional frequencies can also be seen in the data on thiols and amines given by Scott and Crowder[12]. The solvents used in their work were cyclopentane and 2,2,4-trimethylpentane; however, it should be noted that not all solvents show this behaviour. Phenol in benzene solution, for example, shows no evidence for the 310 cm^{-1} torsional frequency.

B. Calculation of the barrier to internal rotation from the observed torsional frequency for phenol

The internal rotation barrier for phenol has been treated in great detail using both microwave[13] and infra-red[11, 14] data. It is well established that the O—H group lies in the plane of the benzene ring and hence the hindering potential can be written in the form

$$V = \tfrac{1}{2}V_2(1 + \cos 2\alpha) + \tfrac{1}{2}V_4(1 + \cos 4\alpha) + \ldots \tag{2}$$

where V_2 and V_4 are the heights of the twofold and fourfold barriers and α is the torsional angle which is $\pi/2$ in the planar configuration. The torsional energy for each torsional quantum number, v, is

$$E_{v,\sigma} = (n/2)^2 F b_{v,\sigma} \tag{3}$$

and the energy solution gives $+$ and $-$ sublevels for each v. In the above equations: v is the principal torsional quantum number, σ is the index designating $+$ and $-$ levels, $b_{v\sigma}$ is an eigenvalue of the Mathieu equation, $F = h^2/8\pi^2 r I_\alpha$, $r I_\alpha$ is the reduced moment of inertia for internal rotation, I_α is the moment of inertia of the internal top (O—H group) about its symmetry axis, n denotes the n-fold barrier (in this case $n = 2$), and $V_4 = \pm 0.01 V_2$ (experimental observation).

The observed infra-red torsional frequency, ω_τ, gives the difference between two of the energy levels, $\Delta E_{v\sigma}$

$$\omega_\tau = (n/2)^2 F(\Delta b_{v\sigma}) = \Delta E_{v\sigma} \tag{4}$$

Thus, if ω_τ, the torsional frequency, is known and there is enough structural information to determine $F\,(=h^2/8\pi^2 r I_\alpha)$, $\Delta b_{v\sigma}$ can be calculated. From $\Delta b_{v\sigma}$, a parameter, s, can be obtained from Mathieu equation solution tables. The barrier height is then

$$V_2 = Fs \tag{5}$$

Thus, if the torsional frequency can be measured accurately, the evaluation of V_2 is straightforward provided I_α can be calculated from known structural information.

The geometry of the O—H rotor relative to the symmetry axis appears to be the only problem in the calculation of I_α. Bist et al.[14], in order to fit their data for both phenol and phenol-OD, were forced to conclude that the oxygen atom assumed an off-axis position

114

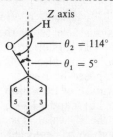

Mathieu et al.[13], through the analysis of new microwave data by rotation–internal rotation theory, find that the frame principal axis is within $\pm 0.18°$ parallel to the internal rotation axis. These authors also point out that the relatively large errors in their microwave constants originate mainly in the limited precision of the infra-red data which in this case were due to Bist et al.[14].

We feel that our present data on phenol will, at least, partially resolve these discrepancies and support the evidence that the C—O rotor axis is coincident with the frame principal axis. These data are given in *Table 3* and compared with the data of Bist et al.[14].

Table 3. Far infra-red spectra of phenol and phenol-OD

Phenol (cm^{-1})		Phenol-OD (cm^{-1})		
Bist[1], vapour	This work, solution	Bist[1], vapour	This work, solution	Assignment/ comment
408	—	382	388	$V_{18b}(b_2)$ x-sens.
309.6	310	—	—	O—H torsion $0 \rightarrow 1$
278.0 276.2	282	—	—	$1 \rightarrow 2$
253.0 237.0	232	—	—	$2 \rightarrow 3$
244*	—	—	—	$V_{11}(b_1)$ x-sens.
—	—	246.4 228.4 228.4 233	247 212	See text

* Raman value—CCl$_4$ solution.

Bist has assigned the 246–228 bands as the $0 \rightarrow 1$ and $1 \rightarrow 2$ transitions of the O—D torsion and the 232 cm^{-1} band as the x-sensitive $V_{11}(b_1)$ mode. These authors also note that their data do not lead to the expected shift of $\sqrt{2}$ in the reduced moment of inertia on going from O—H to OD and ascribe this to a change in geometry. Our solution data indicate that there is a second, intense band at 212 cm^{-1} which was below the range of their instrument. Our data on phenol-OD, as well as data on other phenols to be discussed later, show that the 247 cm^{-1} band is not the $0 \rightarrow 1$ transition for the O—D torsion (b_1) but is one branch of a Fermi doublet arising from resonance between the O—D torsion and the x-sensitive $V_{11}(b_1)$ mode near 233 cm^{-1}. This would then place the O—D torsional frequency near 230

115

cm^{-1} which gives a $v_{O-H}/v_{O-D} = 310/230 = 1.35$ in good agreement with ratios found for the torsions in most of the phenols investigated. This reassignment of the phenol-OD torsion could remove some of the inconsistencies noted by Mathieu[13], and also makes it unnecessary to assume the geometry of the rotor proposed by Bist[14].

Also additional observation of transition between higher energy levels in several phenols, i.e. the hot bands $1 \rightarrow 2$ and $2 \rightarrow 3$, provides adequate information concerning the V_4 value. Usual values of V_4 in the range of $\pm 0.01\ V_2$ are small enough to be neglected in these studies.

III. EXPERIMENTAL PROCEDURES

The various phenol samples were obtained from several sources and were purchased in the highest purity available†. Usually, the samples were used without further purification. Because the torsional band is, in general, much more intense in dilute solution than the other bands in the far infra-red region for these compounds, it is felt that small amounts of impurities would not interfere.

The phenol-OD analogues were prepared by deuterating the OH compound directly in solution with D_2O by a technique described elsewhere[15].

Far infra-red spectra were obtained on a Digilab FTS-14 interferometer system. A 6 μm mylar beamsplitter was used to cover the range 450 to 75 cm^{-1}, and for most samples the resolution was 8 cm^{-1}. This instrument is ideally suited to this type of study because of its ability to ratio out solvent backgrounds even with the low energy available in the far infra-red region.

Cells equipped with 1/16 in. polypropylene windows were employed and the path length was generally 5 mm although for a few samples a 10 mm path length was necessary. A compensating cell was used in the reference beam to ratio out absorption due to the solvent and windows.

IV. RESULTS

The observed torsional frequencies and calculated barriers to internal rotation for several monosubstituted phenols are given in *Table 4*. Also presented in *Table 4* is the difference in the torsional frequency, $\Delta\omega_\tau$, for these compounds compared to the phenol torsion at 310 cm^{-1}. That is

$$\Delta\omega_\tau = \omega_\tau^{obs} - \omega_\tau^{phenol} \qquad (6)$$
$$= \omega_\tau^{obs} - 310\ cm^{-1}$$

These values are presented in *Table 4*.

---

† Commercial samples were purchased from Eastman, Aldrich, Pfaltz and Bauer, and Chem-Service, Inc. We are indebted to Dr A. W. Baker of The Dow Chemical Co., Walnut Creek, California for a number of custom synthesized phenols.

Table 4. Observed torsional frequencies (ω_τ), barriers to internal rotation (V_2, in cm^{-1} and kcal/mol), and $\Delta\omega_\tau$ shift from phenol for various monosubstituted phenols

Phenolic compound	$\omega_\tau,$ cm^{-1}	V_2		$\Delta\omega_\tau,$ cm^{-1}
		cm^{-1}	kcal/mol	
p-Fluoro	280	1034	2.96	−30
-Chloro	302	1185	3.30	− 8
-Bromo	303	1193	3.41	− 7
-Iodo	313	1266	3.50	+ 3
-Methyl	295	1135	3.24	−15
-Methoxy	268	954	2.73	−42
-t-Butyl	301	1178	3.37	− 9
-Phenyl	310	1244	3.56	0
-Nitro	350	1588	4.54	+40
-n-Butoxy	268	954	2.73	−42
-Cyanide	343	1492	4.264	+33
-Methyl	295	1134	3.24	−15
m-Fluoro, *cis*	329	1389	3.97	+19
trans	316	1289	3.68	+ 6
-Chloro	312	1258	3.60	+ 2
-Bromo	312	1258	3.60	+ 2
-Iodo	313	1266	3.62	+ 3
-Methyl	312	1258	3.60	+ 2
-Methoxy	317	1296	3.70	+ 7
-Cyanide	316	1289	3.68	+ 6
-Nitro	320	1318	3.77	+10
-Phenyl	310	1244	3.56	0
-t-Butyl	304	1200	3.43	− 6
-Hydroxy	318	1304	3.73	+ 8
o-Cyanide, bonded*	393	1932	5.52	+83
, unbonded	376	1782	5.09	+67
-Fluoro	366	1694	4.84	+56
-Chloro, bonded*	396	1965	5.62	+87
, unbonded	361	1660	4.74	+52
-Bromo, bonded*	395	1949	5.57	+82
, unbonded	361	1643	4.70	+50
-Iodo, bonded*	378	1826	5.22	+68
, unbonded	345	1526	4.36	+35

* Bonded indicates the intramolecular bond with the halogen (See ref. 16).

Bonded   Unbonded

Now, if the substituents represent linear mesomeric and inductive effects, i.e. each substituent in a specific position *always* influences the π bonding with the same magnitude, then we should be able to use the values in *Table 4* to predict the torsional frequencies and barriers for multisubstituted phenols.

For example, the torsional vibration of the non-bonded 2,4,5-trichlorophenol should be the linear combination of the values in *Table 4*. Summarizing $\omega_\tau^{\text{calc}} = \omega_\tau^{\text{phenol}} + (\Delta\omega_\tau)_{\text{Cl}}^{\text{ortho}} + (\Delta\omega_\tau)_{\text{Cl}}^{\text{para}} + (\Delta\omega_\tau)_{\text{Cl}}^{\text{meta}}$. Substituting the values from *Table 4*, we obtain

$$\omega_\tau^{\text{calc}} = 310\,\text{cm}^{-1} + (+52\,\text{cm}^{-1})_{\text{Cl}}^{\text{ortho}} + (-8\,\text{cm}^{-1})_{\text{Cl}}^{\text{para}}$$
$$+ (+2\,\text{cm}^{-1})_{\text{Cl}}^{\text{meta}} = 356\,\text{cm}^{-1} \qquad (8)$$

The observed torsional frequency of the unbonded 2,4,5-trichlorophenol is $354\,\text{cm}^{-1}$. A summary of this calculation for several phenols is given in *Table 5*. We are very gratified to find the good to excellent agreement between

Table 5. Observed versus calculated torsions

Compound	Torsion			Comment
	Obs.	Calc.	Diff. %	
1. Di-substituted				
2,3-Dimethyl	308	304	−1.3	
3,5-Dimethyl	312	312	0	
2,6-Dimethyl	303	302	−0.3	
3,4-Dimethyl	298	297	−0.3	
3-Methyl-6-chloro	397	397	0	Bonded form
	367	364	−0.8	Free form
3-Methyl-4-chloro	305	304	−0.3	
2,4-Dichloro	389	389	0	Bonded form
	354	354	0	Free form
2-Bromo-4-chloro	387	387	0	Bonded form
	354	352	−0.6	Free form
2-Chloro-4-fluoro	370	367	−0.8	Bonded form
	336	332	−1.2	Free form
2-Fluoro-4-nitro	395	396	+0.3	
3,5-Di-*t*-butyl	304	304	0	
2. Tri-substituted				
2,4,6-Tribromo	392	388	−1.0	Bonded
2,4,6-Trichloro	393	389	−1.0	Bonded
2,4,6-Triiodo	382	383	+0.3	Bonded
2,6-Dichloro-4-bromo	396	388	−2.0	Bonded
2-Br-4,6-Dichloro	393	388	−1.3	Bonded using *o*-chloro value
2,4,5-Trichloro	389	391	+0.5	Bonded
	353	356	+0.8	Free
2,6-Dibromo-4-methyl	388	380	−2.0	Bonded form
2,6-Dichloro-4-nitro	432	437	+1.2	Bonded form
2,6-Dibromo-4-nitro	432	435	+0.7	Bonded form
2,6-Diiodo-4-nitro	417	420	+0.7	Bonded form
2,6-Dichloro-4-hydroxy	360	(360)	0	Predicts *p*-OH shift = −37
2,6-Dimethyl-4-methoxy	254	260	+2.0	254 calc. from OD. Fermi res. in OH compound?
3. Other substituted				
Pentachloro	394	391	−0.8	

calculated and observed torsional frequencies. This confirms our suspicion that these mesomeric and inductive effects are linear and can be predicted for many molecules from the data in *Table 4*.

These values can be used to predict chemical properties much like Hammett's σs have been used in the past. We feel the $\Delta\omega_\tau$ better reflects the aromatic climate of the ring and will give yet another, and we feel better, parameter for the use of the chemist. This can further be extrapolated to predict bond parameters of aromatic rings in polymer chains.

PART I. REFERENCES

[1] W. J. Hehre, L. Radom and J. A. Pople, *J. Amer. Chem. Soc.*, in press.
[2] T. Kojima, *J. Phys. Soc. Japan*, **15**, 284 (1960).
[3] A. Hall and J. L. Wood, unpublished results quoted in G. E. Campagnaro and J. L. Wood, *J. Molec. Structure*, **6**, 117 (1970).
[4] For details see G. L. Carlson, W. G. Fateley and F. F. Bentley, *Spectrochim. Acta*, **28A**, 177–179 (1972).
[5] C. J. Roothaan, *Rev. Mod. Phys.* **23**, 69 (1951).
[6] W. J. Hehre, R. F. Stewart and J. A. Pople, *J. Chem. Phys.* **51**, 2657 (1969).
[7] J. A. Pople and M. S. Gordon, *J. Amer. Chem. Soc.* **89**, 4253 (1967).
[8] R. S. Mulliken, *J. Chem. Phys.* **23**, 1833 (1955).
[9] V. von Keusler, *Z. Elektrochem.* **58**, 136 (1954).
[10] J. C. Evans, *Spectrochim. Acts*, **16**, 1382 (1960).
[11] W. G. Fateley, F. A. Miller and R. E. Witkowski, *Technical Report*, Air Force Materials Laboratory, AFML-TR-66-408, January 1967, Wright-Patterson Air Force Base, Ohio.
[12] D. W. Scott and G. A. Crowder, *J. Molec. Spectrosc.* **26**, 477 (1968).
[13] E. Mathieu, D. Welti, A. Bander and Hs. H. Günthard, *J. Molec. Spectrosc.* **37**, 63 (1971).
[14] H. D. Bist, J. C. D. Brand and D. R. Williams, *J. Molec. Spectrosc.* **24**, 402 (1967).
[15] H. D. Bist, J. C. D. Brand and D. R. Williams, *J. Molec. Spectrosc.* **24**, 413 (1967).
[16] G. L. Carlson, W. G. Fateley, A. S. Manocha and F. F. Bentley, *J. Phys. Chem.* **76**, 1553 (1972).

Part II
Far-Infra-red Investigation of Internal Rotation about Single Bonds (In collaboration with Mr A. S. Manocha and Dr E. Tuazon)

I. INTRODUCTION

In order to understand internal rotation in macromolecules, we must have a better understanding of potentials surrounding such units as carbon–carbon, carbon–oxygen, carbon–nitrogen and carbon–sulphur single bonds. A far infra-red investigation with *emphasis* on the evaluation of the asymmetric potential function should yield the desired barrier characteristic of these units. This part describes this investigation.

II. BACKGROUND

A large body of experimental data on the subject of internal rotation has been accumulated over the past ten years[1]. However, most of these data deal with barrier parameters for a highly symmetrical internal rotor, namely, the methyl group. Much work still remains to be done on molecules containing asymmetric rotors. A number of determinations have been carried out on molecules with asymmetric rotors by microwave spectroscopy[1], but in only a few cases has the potential function been fully characterized[2]. Oftentimes, these measurements are accompanied by large uncertainties in the calculated barrier parameters. These large errors arise when the microwave measurements rely heavily on comparison of intensities between rotational lines in calculating the positions of the different torsional levels. Because of the vibrational nature of the internal rotation (torsion), far-infra-red spectroscopy is the most direct and the more accurate technique for such studies *provided* sufficient transitions between the torsional levels of a molecule are observed.

Increasing efforts are now being exerted on the correlation of existing data with the ultimate objective of explaining the source of the barrier to internal rotation[1a, 3]. In Part I, we described the theoretical calculation of barriers in phenols by Professor Pople's group and the experimental confirmation of these barriers by far-infra-red spectroscopy. This represents a portion of our goal toward understanding barriers. It is along this line that an evaluation of potential functions for asymmetric rotors is of particular merit. It has been pointed out by Lowe[1a] that such results will constitute a more rigorous test for a theory than barrier values for methyl rotors. This is because in theoretical calculations of barriers, the symmetry of the methyl group may force a great deal of error cancellation whereas such errors will still be highly evident for asymmetric rotors.

In both experimental and theoretical determinations, the general Fourier series $V(\alpha) = \sum_n a_n \cos n\alpha + \sum_n b_n \sin n\alpha$ has been widely applied for the potential function of internal rotation. For cases where the potential is an even function of the torsional angle α, the form is reduced to a cosine series only, i.e. $V(\alpha) = \frac{1}{2} \sum_n V_n (1 - \cos n\alpha)$. An interesting suggestion is that the

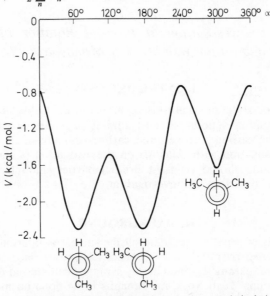

Figure 1. Predicted potential curve for isopropyl alcohol.

120

interaction between an atom or substituent in an internal rotor and the frame may be individually expressed as a Fourier series, each type of interaction being additive. If proven to be valid, such an approach will be very useful in predicting the form of the potential curve for internal rotation. Predictions on this basis were made by Stiefvater and Wilson[2b] for methyl- and fluoro-substituted acetyl fluorides but no experimental confirmation has yet been made.

We have applied the method of Stiefvater and Wilson[2b] to some molecules of interest in this research to see what might be the approximate form of the potential curve associated with internal rotation. Due to lack of any experimental data, we used the theoretical results of Radom, Hehre and Pople[3] on monomethyl- and monofluoro-substituted methanols and methylamines. These authors' results for the internal rotation around the C—O and C—N bonds, using the Fourier cosine form for the potential, are summarized below:

	(kcal/mol)		
Molecule	V_1	V_2	V_3
$CH_3—OH$	0	0	−1.12
$CH_3CH_2—OH$	−0.93	−0.05	−1.14
$FCH_2—OH$	5.25	−2.20	−0.96
$CH_3—NH_2$	0	0	−2.13
$CH_3CH_2—NH_2$	0.89	0.21	−2.29
$FCH_2—NH_2$	−4.86	4.28	−2.01

It is seen that the value of V_3 for each series of molecules given above remains approximately constant, perhaps well within the error due to the uncertainty in the molecular geometries used in the calculation. This near constancy of V_3 has also been observed in the substituted acetyl fluorides[2b]. Thus, in ethanol, the effect of a methyl substitution on the different potential constants can be expressed as follows:

$$V_3^{CH_3/H} \sim 0$$

$$V_1^{CH_3/H} = -0.93$$

$$V_2^{CH_3/H} = -0.05$$

The potential curve for isopropyl alcohol may then be synthesized by superimposing the expansion $[\frac{1}{2}V_1(1 - \cos \alpha) + \frac{1}{2}V_2(1 - \cos 2\alpha)]^{CH_3/H}$ over the expansion $\{\frac{1}{2}V_1[1 - \cos (\alpha + 2\pi/3)] + \frac{1}{2}V_2[1 - \cos 2(\alpha + 2\pi/3)]\}^{CH_3/H}$ and adding the basic V_3 term. The average values of $V_3 = -1.1$ kcal/mol and $V_3 = -2.1$ kcal/mol were assumed for the methanol and methylamine series, respectively. The resulting potential curve for isopropyl alcohol is given in *Figure 1*. Results of similar calculations on isopropyl amine and 1-fluoroethanol are presented in *Figures 2 and 3*.

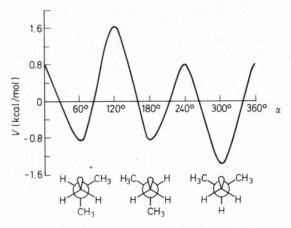

Figure 2. Predicted potential curve for isopropyl amine.

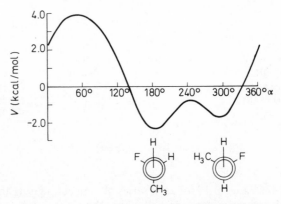

Figure 3. Predicted potential curve for 1-fluoroethanol.

While the above calculations are certainly approximate, experimental verification indicates they are correct. This allows us to anticipate that iso-propyl alcohol and isopropyl amine, like ethanol[4] and ethylamine[5], may exist in the respective *trans* and *gauche* rotamers in measurable proportions. Experimental results on similar larger molecules demonstrated this approximation.

III. EXPERIMENTAL

The far-infra-red spectra of several of the above-mentioned molecules have been measured and the torsional vibrations located for the C—C, C—O, C—N and C—S bonds. As mentioned earlier, as many transitions between torsional levels as possible must be measured for an accurate characterization of the potential curve. The fulfilment of this requirement is favoured by the low-frequency nature of most torsional vibrations.

The spectra will be recorded in this laboratory's newly acquired DIGILAB FTS-14 infra-red Fourier transform spectrophotometer. The maximum resolution capability of this instrument is 0.5 cm^{-1} throughout the region 10 to 4000 cm^{-1}. Interchangeable mylar beam splitters of various thicknesses permit the entire far-infra-red range to be examined. All measurements were carried out in the gas phase to permit higher resolution and accuracy in measuring band centres. Our laboratory is equipped with a heated 1-meter cell, a 10-cm gas cell, and various home-built short-path cells. For resolutions of 1 and 0.5 cm^{-1} in the far-infra-red region, it was found necessary to use wedged polyethylene or polypropylene cell windows to eliminate fringes.

The assignment of the torsional frequencies in C—OH, C—NH$_2$ and C—SH compounds was verified by a solution technique developed in this laboratory[6]. A dilute solute solution of an alcohol or amine in cyclohexane (about one per cent) is made and to this is added a few drops of D$_2$O. The ensuing substitution of the labile hydrogen atom by deuterium permits the measurement of shifts in the torsional frequencies resulting from deuteration.

It is not uncommon that a torsional band is obscured by overlapping with another broad low-frequency band (e.g. methyl torsion) in which case the vapour-phase spectrum of the compound deuterated at the functional position was useful. Such preparation is easily made through exchange reaction in an excess of D$_2$O and separation by distillation. Several deuterated forms of ethanol and ethylamine were purchased for this study.

IV. INTERPRETATION OF DATA

For the molecules of interest here, the Fourier cosine series, $V(\alpha) = \frac{1}{2}\sum_n V_n$ $(1 - \cos n\alpha)$, was adequate to represent the potential associated with internal rotation. It has been found that, in general, the series converges rapidly with $n = 3$ or 4 being sufficient to explain the spectroscopic data. The wave equation for the internal rotation problem is a Mathieu-type equation in one dimension

$$[p_\alpha F(\alpha)p_\alpha + V(\alpha)]M(\alpha) = EM(\alpha)$$

where $F(\alpha)$ is the reduced moment of inertia parameter, p_α is the momentum conjugate to α, E is the energy eigenvalue, and $M(\alpha)$ is a function expressed as an expansion in some chosen basis set.

A computer programme has been written[7] which utilizes the free-rotor functions $\exp im\alpha$ as the basis set for calculating the matrix elements, i.e. $M_n(\alpha) = \sum_n a_{nm} \exp im\alpha$. The size of the basis set can be varied and it is determined through the convergence of eigenvalues at a particular energy level. The variation of the inertial parameter F with the angle of internal rotation has been included by incorporating a programme written by Meakin et al.[2a] for such purpose. The complete programme has been tested with the known data on molecules with asymmetric rotors like H$_2$O$_2$[8], as well as on those containing symmetrical rotors (e.g. methyl group). The procedure consists of making a judicious choice of an initial set of potential constants V_n and comparing the calculated eigenvalue differences with the observed frequencies. The values of V_n are then refined to give the best fit to the observed data through an iterative numerical procedure.

We have carried out a zero-order calculation on ethanol using the theoretical values of V_1, V_2 and V_3 given by Radom, Hehre and Pople[3]. The small variation of the reduced moment of inertia of the OH group with the internal angle was neglected; thus, a constant value $F = 21$ cm^{-1} was calculated from an approximate geometry. The predicted energy levels for ethanol are depicted in *Figure 4* (n is just a number that gives the order of the levels). The first

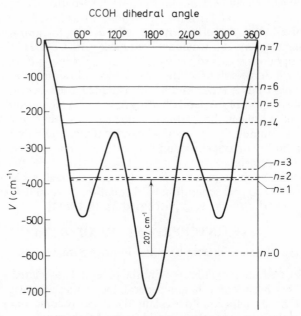

Figure 4. Predicted energy levels for ethanol.

two levels, assigned to the *trans* form, have an energy separation of 207 cm^{-1}. Lake and Thompson[9] observed the torsional band of ethanol in the vapour phase under low resolution and reported the centre of the broad band at 199 cm^{-1}. The measured band contour indicates that, under higher resolution, the transitions between the upper torsional levels might be observed. These higher transitions allow a more accurate determination of the potential constants.

V. THE EFFECT OF HIGHER TERMS AND VARIATION OF THE INERTIAL PARAMETER ON THE THREEFOLD POTENTIAL FUNCTION

In most determinations of the threefold barrier V_3 for the methyl groups, the next higher potential constant V_6 in the cosine series expansion has been neglected. While the values of V_6 have indeed been found to be generally small (about one per cent of V_3), their accurate determination, particularly their sign, will be useful in distinguishing between various theories on the source of barrier in methyl rotors[1a]. In their semiempirical electrostatic model for internal rotation, Lowe and Parr[10] pointed out that a negative V_6 is connected with those charges and multipoles (situated in the plane containing the heavy atoms) which repel methyl protons, while a positive V_6 is produced by those charges and multipoles (in the plane) which attract the methyl group.

In nearly all calculations on threefold potential, the variations of the reduced moment of inertia of the top (F) during the torsional motion, such as those which arise from bond stretching and angle deformation, have been neglected. Recently, Ewig and Harris[11] demonstrated that such assumption of constant results in an apparent V_6 component in the potential even if none, in fact, existed. These authors also emphasized that it is not possible to distinguish between the effect of a variation in the reduced moment of inertia of the top and contribution of higher terms to the threefold potential.

Harris *et al.*[12] showed that the Hamiltonian $H(\alpha) = p_\alpha F(\alpha) p_\alpha + V(\alpha)$, expressed in the geometry related coordinate α, may be transformed to the form $H(\theta) = p_\theta F_0 p_\theta + V'(\theta)$ where θ is a coordinate which removes the angle dependence of F. The coordinate θ has no attachable geometrical meaning and only follows the condition that it be equal to α at 0 and 2π radians. The two forms of the Hamiltonian give the same eigenvalues. The transformation from α to θ produces a distortion relative to the coordinate (shifts in the position of maxima and minima and change in shape of the potential curve) but does not affect the height of the barrier. In the special case of the methyl rotor, symmetry dictates that α and θ be the same at the positions of maxima and minima and the only question that arises is how both F and V_6 augment the shape of the potential curve.

We must note in applying these theories to macromolecules that some attention be directed to the study of F and V_6 dependence of the threefold potential. Several molecules with methyl rotors and with well-determined geometry in the ground vibrational state were chosen for this evaluation. Their far infra-red spectra were recorded with the object of measuring as many transitions between torsional levels as possible. The task of making the choice of these molecules is made easier by earlier measurements in this laboratory[13]. With better instruments available to us now, improved spectra in terms of resolution and frequency accuracy show these splittings of energy levels.

Although there is an apparent inseparability of the F and V_6 effects, much information has been gained by the following simple approach. In the *first* step, a fit of the observed torsional transitions to only F_0 and $V'_3 (= V_3)$ is attempted. Even though only two frequencies are needed to determine F_0 and V'_3, it must be required that two or more additional transitions be provided to serve as internal check. The calculated F_0 will then be referred to the value of F initially computed from the geometry in the ground vibrational state. The difference $(F_0 - F)$ can then be related to the change in molecular geometry during the torsional motion. In molecules with relatively heavy frames, it is perhaps a good assumption to consider the change as occurring mainly in the methyl group through some C—H elongation or HCH angle deformation or both. When such calculated changes are reasonably well within expected limits, a successful fit to only F_0 and V'_3 will mean that V'_6 (or V_6 itself) is most likely zero.

125

In the *second* step, when the geometry related change $(F_0 - F)$ is unreasonably large for a fit to only F_0 and V'_3 or when no such fit can be successful, the same iteration technique will be applied using F_0, V'_3 and V'_6. Whether this operational determination of F_0 and V'_6 gives the correct contribution of each parameter to the potential is difficult to determine. It is hoped that with enough test cases studied a criterion can be reached as to the extent to which each factor contributes in macromolecular molecules.

VI. SUMMARY

When put on a firmer foundation, the subject of internal rotation and unbonded forces will undoubtedly have far-reaching implications for understanding the intricate structure of simple molecules which have stable rotamer forms and be of use in understanding macromolecular and biological molecules. At present, the sizable amount of data accumulated on the subject is strictly for the consumption and use of theoretical chemists who are continuously looking for an elegant theory that will explain the origin of the barrier to internal rotation. Our efforts are directed toward understanding conformations in macromolecules.

PART II. REFERENCES

[1] (a) J. P. Lowe, *Progr. Phys. Org. Chem.* **6**, 1 (1968);
(b) Y. Morino and E. Hirota, *Annu. Rev. Phys. Chem.* **20**, 139 (1969);
(c) H. D. Rudolph, *Annu. Rev. Phys. Chem.* **21**, 73 (1970).
[2] (a) P. Meakin, D. O. Harris and E. Hirota, *J. Chem. Phys.* **51**, 3775 (1969);
(b) O. L. Stiefvater and E. B. Wilson, *J. Chem. Phys.* **50**, 5385 (1969);
(c) E. Saegebarth and E. B. Wilson, *J. Chem. Phys.* **46**, 3088 (1967);
(d) E. Hirota, *J. Molec. Spectrosc.* **26**, 335 (1968);
(e) E. Hirota, *J. Chem. Phys.* **37**, 283 (1962).
[3] L. Radom, W. J. Hehre and J. A. Pople, to be published.
[4] (a) Y. Sasada, M. Takano and T. Satoh, *J. Molec. Spectrosc.* **38**, 33 (1971);
(b) M. Takano, Y. Sasada and T. Satoh, *J. Molec. Spectrosc.* **26**, 157 (1971).
[5] (a) Y. S. Li and V. W. Laurie, *Symp. Molec. Struct.* paper 01 (1969);
(b) T. Masamichi, A. Y. Hirakawa and K. Tamagake, *Nippon Kagaku Zasshi*, **89**, 821 (1968).
[6] G. L. Carlson, W. G. Fateley and F. F. Bentley, *Spectrochim. Acta*, **28A**, 177–179 (1972).
[7] Graduate project of A. S. Manocha supported by Carnegie–Mellon University (1971).
[8] R. H. Hunt, R. A. Leacock, C. W. Peters and K. T. Hecht, *J. Chem. Phys.* **42**, 1931 (1965).
[9] R. F. Lake and H. W. Thompson, *Proc. Roy. Soc. A*, **291**, 469 (1966).
[10] (a) J. P. Lowe and R. G. Parr, *J. Chem. Phys.* **44**, 3001 (1966);
(b) J. P. Lowe, *J. Chem. Phys.* **45**, 3059 (1966).
[11] C. S. Ewig and D. O. Harris, *J. Chem. Phys.* **52**, 6268 (1970).
[12] D. O. Harris, H. W. Harrington, A. C. Luntz and W. D. Gwinn, *J. Chem. Phys.* **44**, 3467 (1966).
[13] W. G. Fateley and F. A. Miller, *Spectrochim. Acta*, **19**, 611 (1963).

THE ROLE OF ENERGY TRANSFER IN THE STABILIZATION OF POLYMERS

J. E. GUILLET

Department of Chemistry, University of Toronto, Toronto 181, Canada

ABSTRACT

Energy transfer processes can, in principle, represent a powerful new mechanism for the stabilization of polymers against degradation by ultra-violet light. The mechanism involves the transfer of electronic excitation from an absorbing group in the polymer to a 'quenching' group by either an intra- or intermolecular process. The efficiency of such processes depends in a complex way on the nature of the excited states of the donor and acceptor groups, and on the mobility of the polymer chain in the solid state. Studies are reported on the quenching of the photodegradation of polymers containing ketone and aromatic groups by various stabilizers.

INTRODUCTION

The phenomenon of 'weathering' of plastic materials usually consists of a complex sequence of chemical reactions, often initiated by the u.v. radiation of the sun. Solar radiation consists of a nearly Boltzmann distribution of energy with wavelengths extending from about 300 nm in the ultra-violet up to the infra-red region of the spectrum. There is a rather sharp cut-off at about 300 nm because the ozone layer of the atmosphere filters out the short wavelength radiation.

Figure 1 shows such a distribution curve in which the wavelength of the light has been converted to the energy per quantum in kilocalories per mole. The bond strengths of different kinds of chemical bonds that one might expect to find in polymers is also included. Only a very small amount of the total radiation is sufficiently energetic to break strong bonds such as NH and CH, but about ten per cent is sufficiently energetic to break a carbon–carbon bond. However, there is plenty of radiation in the sun's spectrum of sufficient energy to break the weak bonds, such as O—O, N—N and C—Cl. In simple terms this means that if you want to have a stable polymer, you should introduce into the polymer only stable bonds which will reduce the probability of degradative processes occurring on the absorption of light.

POLYMER PHOTOCHEMISTRY

Most commercial polymers when pure do not usually absorb in the spectral region above 300 nm and hence should not degrade in the ultra-

127

violet light of the sun. The first law of photochemistry states that there cannot be a photochemical reaction if there is no absorption. However, processed polymers do in fact have a slight absorption which extends up into the region above 300 nm. The usual mechanism for nearly all polymers which degrade in ultra-violet light is a sensitized photo-oxidation which is induced by small amounts of groups which absorb in this region of the spectrum. In our studies of photochemistry we have attempted to introduce into ordinary polymers such as poly(styrene) and poly(methyl methacrylate) controlled amounts of chemical groups absorbing in this region of the spectrum to study their photochemical reactions. The most common group which exists in polymers and which causes such photochemical reactions is the ketone group.

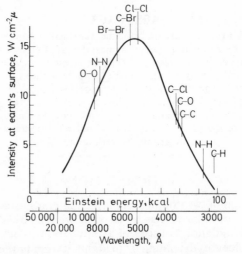

Figure 1. Distribution of solar energy.

In any photochemical reaction, the ways in which the energy of an absorbed photon is dissipated can be classified into photophysical processes and photochemical processes. The photophysical processes include fluorescence, which is simply re-emission of a quantum of light after absorption; phosphorescence, which is emission at a slightly longer wavelength and a slightly greater time interval; conversion to thermal energy; and energy transfer. The alternatives to these photophysical processes, the processes which compete with them for the energy of the quantum absorbed, are photochemical reactions such as the formation of free radicals, intermolecular rearrangements, cyclization, photoelimination and photoionization. Nearly all of these chemical processes will ultimately cause a deterioration in the physical properties of a plastic. It is true that if a plastic is crosslinked, in some cases initial improvement of its properties will occur. But if this process continues extensively, degradation will ultimately take place. When stabilizing a polymer, it is therefore desirable to eliminate the photochemical reactions, i.e. to maximize the photophysical processes, particularly the conversion to thermal energy, and to minimize the photochemical processes. If one can increase the efficiency of this process, a much more stable polymer will result.

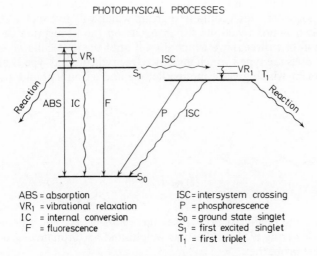

PHOTOPHYSICAL PROCESSES

ABS = absorption
VR_1 = vibrational relaxation
IC = internal conversion
F = fluorescence

ISC = intersystem crossing
P = phosphorescence
S_0 = ground state singlet
S_1 = first excited singlet
T_1 = first triplet

Figure 2. Modified Jablonski diagram.

Figure 2 shows an energy diagram which describes the kinds of processes which are possible. When a quantum of light is absorbed, a particular group is raised to some excited state, and since most groups will be in a singlet ground state, they will be excited to a vibrationally excited singlet state. In the solid phase there will be a rapid degradation of the vibrational energy down to the lowest vibrational level of the electronically excited singlet state, indicated as S_1. From this S_1 excited state, all the processes that we are concerned with will occur. The simplest is fluorescence, that is emitting by dropping down to the ground state singlet. The second is internal conversion, that is conversion of that electronic energy into thermal energy. It is a spectroscopically forbidden process but there are various means by which it can occur and in many polymers it is the major process. The spin of the excited electron can also be inverted to give the triplet state, by intersystem crossing to a vibrationally excited triplet and dropping down to the lowest vibrational state, T_1, of the triplet. Chemical reaction can occur from either the singlet or the triplet excited state. Similarly, by intersystem crossing one can convert the triplet energy to thermal energy, or a quantum of light can be emitted by phosphorescence.

KETONE PHOTOCHEMISTRY

In our laboratory we have studied these processes involving ketones included either in the backbone or the sidechain of a polymer. These ketone groups are important because they are among the few functional groups

129

which do absorb in this region above 300 nm and they are byproducts of the photo-oxidation of polymers. They seem to be the photosensitive groups involved in the fundamental degradation processes in many polymer systems. If a ketone is incorporated in the backbone of the polymer (such as by copolymerizing ethylene and carbon monoxide, for example, or by making a polyester which contains a group which will give such a structure), there are two major reactions which occur on the absorption of light[1, 2]. (There are in fact three more which do not alter the molecular weight.) The two major ones are type I which gives two free radicals and type II molecular rearrangement which gives a methyl ketone and a double bond and results

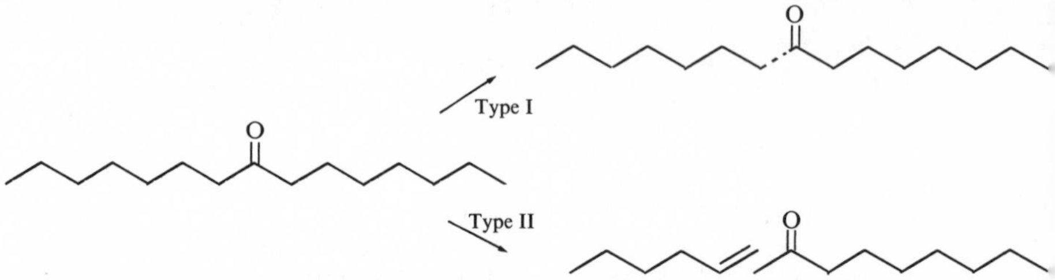

Type I

Type II

in breaking the chain. In both cases the chain breaks and this ultimately leads to a lowering of the molecular weight and an ultimate degradation in the physical properties.

The ketone groups may also be contained in sidechains of the polymer[3, 4]. Upon photolysis, the sidechain ketones indicated by this structure also

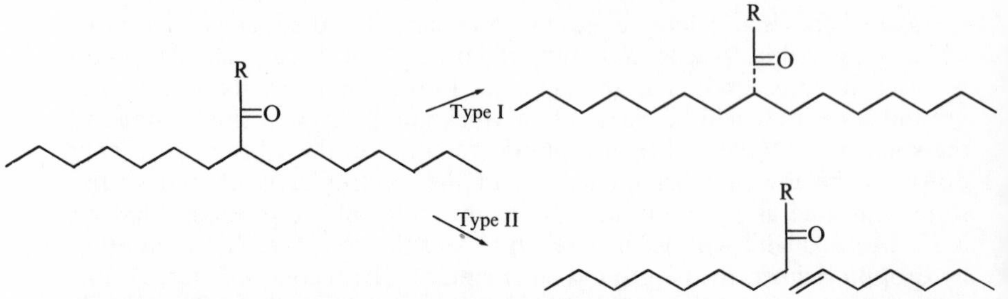

Type I

Type II

undergo the type I and type II reactions. Type I in this case does not give an actual break in the main chain. Type II is the only reaction in this case where the molecular weight of the polymer is reduced and consequently with sidechain ketones this is the mechanism by which the reduction in molecular weight takes place. Since most of the oxidation processes by which ketone structures are introduced into polymers will give sidechain ketones, this appears to be a very important reaction in the photodegradation of polymer systems.

Usually an u.v. absorber is added to the polymer to stabilize such systems. Additives, such as substituted 2-hydroxy benzophenones, have molar extinction coefficients about a thousand times greater than that of the ketone group in the polymer. Consequently a very small amount of this material added will absorb 90 or 95 per cent of the light and prevent the light from being absorbed by the ketone carbonyl. This is a simple but very effective mechanism for stabilization, and it is the major mechanism by which stabilization of polymers has been achieved in the past. The only alternative to this kind of stabilization mechanism is to remove the energy absorbed before chemical reaction can occur. The most effective mechanism for doing this is to transfer the energy from the site at which it has been absorbed in the polymer to some other location. This process we call 'energy transfer'.

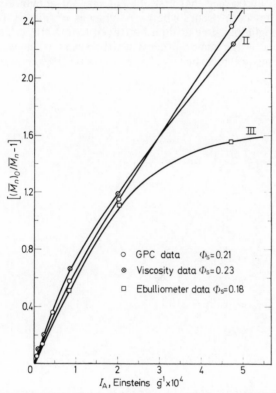

Figure 3. Photodegradation rate of MMA–MVK copolymer (3% MVK) determined by gel permeation chromatography, intrinsic viscosity, and ebulliometric methods. The copolymer was irradiated in benzene solution at 25°C, 313 nm.

The efficiency of a photolytic process is defined in terms of the quantum yield. The quantum yield for scission, Φ_s, is measured simply by measuring the molecular weight of the polymer by various means, and plotting

$$((\overline{M}_n)_0/\overline{M}_n - 1)$$

as a function of the light absorbed. Such a graph is shown in *Figure 3* using three different molecular weight measurements: gel permeation chromatography, viscosity, and ebulliometry. The initial slope of these curves is identical in the three cases and the chain scission quantum yield here is about 0.20 for a methyl vinyl ketone–methyl methacrylate copolymer in solution. This quantum yield means that 20 per cent of the quanta absorbed

131

by the polymer result in an actual break in the polymer chain. If the process were perfectly efficient, the quantum yield would be equal to one.

LUMINESCENCE OF POLYMER SYSTEMS

If one looks at the luminescence of a polymer system as a function of temperature, it can be seen that first of all at very low temperatures we find that only luminescence occurs when a polymer is irradiated. Degradation processes are negligibly slow at liquid nitrogen temperatures[5]. At $-180°C$ phosphorescence is the major process which occurs when we irradiate a ketone-containing polymer such as a styrene–methyl vinyl ketone copolymer.

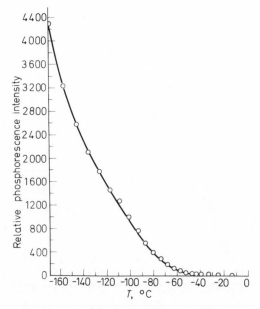

Figure 4. Phosphorescence intensity as a function of temperature, PS–MVK film.

(See *Figure 4.*) The yield of phosphorescence drops off drastically as the temperature increases towards room temperature. At room temperature there is practically no visible phosphorescence. Chemical reaction in all cases requires the movement of atoms and molecular groups within the polymer structure. At very low temperatures the rigidity of the structure is such that these movements cannot occur. As a result, the photophysical processes are nearly 100 per cent efficient. The energy that is absorbed originally as the singlet is converted to the triplet and the triplet emits a quantum of light which we call phosphorescence.

Fluorescence, on the other hand, is almost independent of temperature, as shown in *Figure 5.* The yield is much lower than the phosphorescence at all temperatures and nearly constant with temperature. For polymers at very low temperatures the major process involved will be the conversion of

132

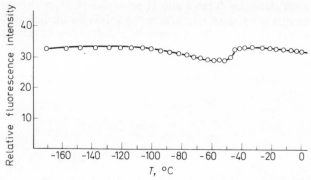

Figure 5. Fluorescence intensity as a function of temperature, PE–CO film.

energy, first of all from the singlet to the triplet by intersystem crossing, and from the triplet to the ground state by phosphorescence. As the temperature is increased, the mobility of the polymer· becomes much greater and the phosphorescence decreases so that more of the energy is converted either into thermal energy or into chemical reactions.

At temperatures less than − 100°C we can expect that nearly all the energy will go into luminescence[6]. If there is an easy conversion into a triplet, phosphorescence is the favoured process. In the range − 100°C up to the glass transition temperature, the major process involved will be the conversion of electronic excitation energy into thermal energy, and at temperatures greater than T_g (the melting point of the polymer) there are two important processes, photochemical reactions and thermal conversion. Below T_g or T_m the photochemical quantum yields are quite small, but still sufficiently large to cause degradation over a long period of time.

PHOTOCHEMISTRY IN THE SOLID STATE

One of the easiest ways to study polymers is to put them in solution and study them in the liquid state. However, since most plastics are used in the solid state, one wants to know what happens under these conditions. It was shown some years ago[1] that in poly(ethylene-CO) the quantum yields for various processes varied with temperature in solution. The type I quantum yield, which is quite small at ordinary temperatures, increases to the equivalent of the type II at 90°C. The quantum yield for luminescence is extremely small at both these temperatures, and the quantum yield for conversion of the electronic energy to thermal energy is quite large in both cases, namely 0.97 and 0.95. For poly(ethylene) at ambient temperatures the major process is still conversion to thermal energy. Only a very small portion of the total energy ends up in the degradation process. However, it is this small portion that we would like to get rid of because these polymers in fact degrade very rapidly in ultra-violet light. There is a very large u.v. component in the sun's radiation and a process which is only two per cent efficient can still cause a great deal of degradation.

The quantum yields for types I and II processes in poly(ethylene-CO) at room temperature are almost identical in the solid and the liquid state. This is because poly(ethylene) at this temperature is well above its glass transition ($T_g - 100°C$) and thus the molecules in poly(ethylene) are sufficiently mobile so that all of the photochemical processes (i.e. all the rearrangements of the molecular structure which are necessary to carry out the reaction) can take place.

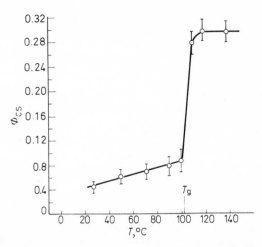

Figure 6. Quantum yield of chain scission as a function of temperature, PS–PVK film.

In polymers such as poly(styrene) which have glass transitions in the region of interest, there is a strong effect of the glass transition on quantum yields. For example, in (styrene–phenyl vinyl ketone) the quantum yield for chain scission increases slightly with temperature up to the glass transition[5] (Figure 6). Below T_g, Φ_s is still fairly low, 0.04 to 0.08, but there is a sudden drastic increase at the glass transition, going up to about 0.3, which is exactly the quantum yield for the same process in the liquid state.

When the phosphorescence of such a copolymer is studied as a function of temperature, one will see the effect of molecular mobility. If one plots the logarithm of the intensity of phosphorescence as a function of $1/T$, one finds that two straight lines are obtained. For a methyl methacrylate copolymer containing a small amount of methyl vinyl ketone two lines are obtained corresponding to two activation energies. The transition at $-140°C$ corresponds to a well-known transition at which motion of the α-methyl group occurs. Rotation of these groups increases the probability of energy transfer and of conversion of electronic excitation into thermal energy. This then represents a more fundamental aspect of photochemistry. If one wishes to convert more energy to thermal energy, that is to stabilize the polymer, it is necessary to have molecular motion of the pendant groups in the polymer.

In poly(methyl methacrylate) the α-methyl group rotates at $-140°C$, and the ester methyl group at $-220°C$. The whole ester group rotates at $-20°C$ and of course the glass transition corresponds to the motion of long segments of the chain at $+100°C$. Photochemical evidence for all of these processes can be found. All of these various forms of motion of the polymer will affect the efficiency of the photochemical process.

ENERGY TRANSFER

Energy is transferred in a number of ways in a ketone system. The initial excitation occurs in the carbonyl group, but reaction actually takes place at either the α–β or β–γ C—C bonds, so that there must be even in photo-chemical reaction a transfer of the energy from one bond system to another. That is a very simple case of energy transfer. However, the more practical cases are those in which we can introduce groups into the polymer itself, either as part of the polymer chain or as an additive which will remove the energy of the absorbed quantum during the lifetime of the excited state of such a molecule. There are various kinds of energy transfer processes. The so-called 'Förster' mechanism involves the long range transfer of energy across quite long distances in the polymer, up to 100 Å due to dipole–dipole interactions. Short range transfer involves the so-called 'exchange mechan-ism' which is simply a transfer of electronic excitation between an excited molecule and a non-excited molecule during a collision, due to overlap of electronic charge clouds. In addition, there are exciton mechanisms which occur in crystals and also in polymers, and cause delocalization of energy over a large distance and this results in a transfer of energy from one position to another.

There are two models which are of particular interest in stabilizing polymers. The first is the so-called Stern–Volmer model which involves a dynamic process similar to diffusion. The equation for this is

$$\Phi_0/\Phi = 1 + k_q\tau[Q]$$

where $[Q]$ is the concentration of the quenching molecule, k_q is the rate constant for collision, and τ is the lifetime of the excited state. In many cases we can equate k_q, the collisional rate constant, with k_d, the rate constant for diffusion. In some cases, if we are dealing with a rigid polymer system, in particular with polymer glasses at low temperatures, we will use the so-called Perrin model. This model implies that quenching will occur if a quencher is within a certain distance of the excited molecule. This gives an exponential concentration term in the quencher of this form

$$\Phi^0/\Phi = \exp(-vN[Q])$$

where $v = \frac{4}{3}\pi R_0^3$ and N is Avogadro's number. *Figure 7* shows an experi-mental plot of a Stern–Volmer equation for a poly(ethylene)–carbon mon-oxide copolymer containing ketone groups, with cyclo-octadiene which is a quencher which does not absorb any light of the wavelength used[7]. Instead of getting a straight line as predicted from the Stern–Volmer equation, the line is originally straight and then curves over. This is because the cyclo-octadiene quenches only the triplet state, whereas reaction out of the singlet

135

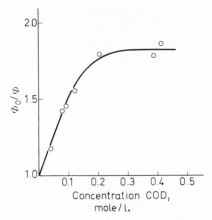

Figure 7. Quenching of chain scission in poly(ethylene–CO) by 1–3 cyclo-octadiene.

groups, poly(phenyl vinyl ketone) in this case (*Figure 8*), the line is quite is not quenched. From such a curve one can determine the portion of the reaction which comes from the singlet and the portion which comes from the triplet. With aliphatic ketones, both the singlet and the triplet reactions are important. On the other hand, when dealing with aromatic ketone groups, poly(phenyl vinyl ketone) in this case (*Figure 8*), the line is quite

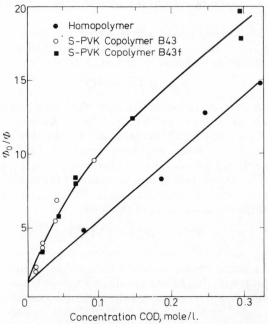

Figure 8. Stern–Volmer plot for quenching of photoscission in poly(PVK) and S–PVK co-polymers by COD in benzene solution.

straight, right up to the highest concentrations and, in fact, the cyclo-octadiene will inhibit at least 99 per cent of the total reaction. With an aromatic ketone we know that internal conversion is almost complete in these systems. Almost all of the reaction comes out of the triplet state and the triplet quencher will be almost totally effective in removing the excited state energy before reaction can take place. In the copolymer there is a curvature which is due to the fact that two kinds of ketone groups are present in the copolymer: isolated ketone groups (i.e. ketones surrounded by styrene units) and sequences of ketone groups. These have different lifetimes and quantum yields as shown in *Table 1*.

Table 1. Comparison of copolymer and homopolymer

Type	Φ_0	$k_q\tau$	τ (sec $\times$ 10^8)
Homopolymer	0.245	43	1.7
Copolymer	0.176	120	4.8

Synthetic polymers are chemically unique because in general they consist of a carbon backbone with chromophoric or other groups regularly spaced along the sidechain of the polymer. Poly(methyl vinyl ketone), for example, has a structure of this kind

This means that there is a sequence of chromophores which are closely held at a similar distance from each other. In fact one can almost think of this as being a linear crystal—a one-dimensional crystal in which there is a fixed relationship between the distances of these carbonyl groups from each other. No matter how one dilutes a polymer or stretches it, this relationship will hold. If light is absorbed in such a molecule, the excitation may be delocalized over several of these groups. In fact there is now very good evidence that in some cases excited state energy can hop from one of these to another along the chain. This has a number of effects on various photochemical processes. For example, in poly(methyl vinyl ketone) the quantum yield for degradation is lower by a factor of ten than if the ketone groups are isolated as in a copolymer, possibly because of the delocalization of the excitation energy[4].

Figure 9 shows the absorption and fluorescence spectrum of poly(methyl isopropenyl ketone) and poly(methyl vinyl ketone) as compared with acetone and heptanone[8]. In absorption, the absorbance of the polymeric ketones is greater and the increase in molar extinction coefficient is an indication of the interaction of the ketone chromophore groups along the chain. In addition, the fluorescence wavelengths are shifted, as is the peak maximum for the absorption.

If ketones are used to quench naphthalene fluorescence, one finds that the polymeric ketones are much less efficient than heptanone or acetone. If, on

137

the other hand, the luminescence from the ketone is quenched by, say, biacetyl, one finds quite the opposite effect, that is the efficiency factor for the small molecules is considerably less than it is for the large polymeric molecules. Thus a polymeric molecule exhibits a low efficiency for the quenching of naphthalene but a high efficiency for the molecule being quenched itself. This is evidence for the delocalization of the energy across the long polymer chain.

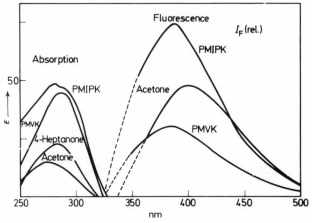

Figure 9. Absorption and emission spectra of various ketones in solution.

There is one other mechanism of energy transfer, the transfer of triplet energy in naphthalene groups. This was shown by Fox and Cozzens[9] with poly(vinyl naphthalene) at cryogenic temperatures and is a process whereby because of the long lifetime of the naphthalene triplet, absorption of two photons in a polymer takes place during the lifetime of one of the excited states. There is a finite probability of having two excited states on the same polymer molecule at the same time. These excited states can migrate by an energy transfer process, either by hopping or by a long range process, and when the two triplets meet, delayed fluorescence results. That is, the two triplets combine to give a singlet excited state which fluoresces and a ground state molecule. This gives emission at fluorescence wavelengths, but with the lifetime of phosphorescence.

We have carried out similar studies with poly(naphthyl methacrylate)[10] and the absorption spectra for normal fluorescence, delayed fluorescence and phosphorescence were observed (*Figure 10*). The delayed fluorescence has the same lifetime as phosphorescence and can be observed in a phosphorescence accessory. The intensity of delayed fluorescence plotted versus that of the phosphorescence gives a straight line with a slope of two, which is precisely the kinetics one would predict in order to have a bimolecular quenching process.

At 25°C the fluorescence spectrum of poly(naphthyl methacrylate) depends on the solvent (*Figure 11*). In a good solvent such as chloroform, only normal

138

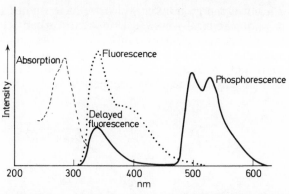

Figure 10. Absorption and emission spectra of poly(α-naphthyl methacrylate).

fluorescence occurs. A poor solvent gives excimer fluorescence which is caused by the interaction of one excited with one ground state naphthalene unit. In the bulk phase only the excimer fluorescence is observed. In mixed solvents, however, the spectrum is removed from the normal fluorescence to the excimer fluorescence. We can demonstrate that it is not possible to form the excimer between two adjacent naphthalene groups. Consequently the polymer molecule then must bend back on itself and form the excimer. In a good solvent the chain is expanded so the probability of forming the excimer is very small. On the other hand, in a poor solvent, the chains are tightly coiled up on each other just as they are in the bulk polymer and there is a very high probability of forming the excimer. In the bulk and in the poor solvent, the excimer is formed and fluoresces. In a good solvent, the excimer does not form to any appreciable extent and normal fluorescence occurs.

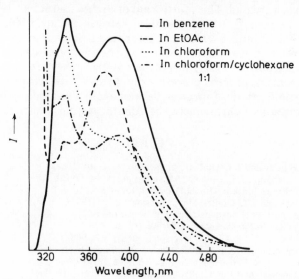

Figure 11. Fluorescence of poly(α-naphthyl methacrylate) in various solvents.

One other form of energy transfer which we have studied is the energy transfer from photochemically excited ketones to carbon tetrachloride[11]. The importance of this form of energy transfer lies in the fact that when we put plasticizers and other molecules into polymer systems, quite different photochemistry can result. A ketone dissolved in carbon tetrachloride gives primarily the photochemical reaction products of CCl_4. The excited state

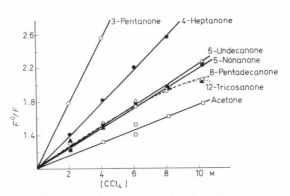

Figure 12. Quenching of ketone fluorescence by carbon tetrachloride.

of the ketone forms what is called an 'exciplex', a complex with carbon tetrachloride, and the carbon tetrachloride then becomes excited, the ketone becomes deactivated, and the decomposition products of carbon tetrachloride result, usually chlorine atoms and $\cdot CCl_3$ radicals. *Figure 12* shows that CCl_4 quenches the fluorescence of the ketones added to it. As the fluorescence is quenched by the carbon tetrachloride the amount of chemical reaction is increased. This fourth kind of energy transfer process is due to an excited state charge transfer complex between the absorbing molecule (the ketone) and another molecule, in this case carbon tetrachloride.

In summary, many types of energy transfer processes, both inter- and intra-molecular can occur in polymer systems. A more thorough study of the mechanisms of these processes should assist in devising practical routes for the stabilization of macromolecules against weathering damage due to the absorption of ultra-violet light from the sun.

REFERENCES

[1] G. H. Hartley and J. E. Guillet, *Macromolecules*, **1**, 165 (1968).
[2] P. I. Plooard and J. E. Guillet, *Macromolecules*, **5**, 405 (1972).
[3] F. J. Golemba and J. E. Guillet, *Macromolecules*, **5**, 212 (1972).
[4] Y. Amerik and J. E. Guillet, *Macromolecules*, **4**, 375 (1971).
[5] E. Dan and J. E. Guillet, manuscript in preparation.
[6] J. E. Guillet, *Naturwissenschaften*, **59**, 503 (1972).
[7] M. Heskins and J. E. Guillet, *Macromolecules*, **1**, 97 (1968).
[8] A. C. Somersall and J. E. Guillet, *Macromolecules*, **5**, 410 (1972).
[9] R. B. Fox and R. F. Cozzens, *Macromolecules*, **2**, 181 (1969).
[10] A. C. Somersall and J. E. Guillet, *Macromolecules*, **6**, 218 1(973).
[11] J. O. Pavlik, P. I. Plooard, A. C. Somersall and J. E. Guillet, *Canad. J. Chem.* **51**, 1435 (1973).

SOME ASPECTS OF STABILIZATION OF POLYMERS AGAINST LIGHT

H. J. HELLER and H. R. BLATTMANN

CIBA-GEIGY Limited, CH-4002 Basel/Switzerland

ABSTRACT

Different mechanisms for the protection of polymers against photodegradation are discussed. (1) The u.v.-absorbers of the 2-(2-hydroxyphenyl)-benzotriazole and the o-hydroxybenzophenone type have rates of internal conversion much higher than the rate of intersystem crossing and of fluorescence decay. Evidence is presented that this rapid non-radiative deactivation cannot be explained very satisfactorily by a simple 'enol–keto'-equilibrium in the first excited singlet state only.

(2) At the concentrations used in practice the quenching rate of light stabilizers is so low that quenching cannot compete successfully with other deactivation processes of triplet states, while excited singlet states of primary sensitizers can be quenched efficiently only by additives with extinction coefficients of more than 5000 to 10000.

(3) The derivatives of 2,2,6,6-tetramethyl-piperidines of the structures I to IV cannot quench excited singlet states in the apolar solvent heptane. The corresponding N-oxyls inhibit the Norrish reaction of aralkylketones. Specific N-oxyl- and N-methyl-2,2,6,6-tetramethyl-piperidines as well as certain nickel chelates decelerate the singlet oxygen induced photo-oxygenation of rubrene and 9,10-dimethoxy anthracene.

1. INTRODUCTION

The effect of light on today's bulk polymers such as the commercially available polyolefins, aliphatic polyamides, linear polyesters, polystyrene, PVC and unsaturated polyester resins is highly complex. At least in their non-light stabilized form these polymers are degraded upon light exposure much less by chainbreaks due to direct photon impact than by normal autoxidation induced photochemically. First of all high energy u.v. light below 280 to 290 nm is completely missing from daylight. Consequently a large proportion of known photodissociations of polymers with aliphatic backbones cannot take place under normal ageing conditions. Secondly, the absorbance of these polymers in the near u.v.-region is rather low due to the lack or low concentration of chromophores of high absorptivity. The photochemistry of everyday polymers depends, therefore, on low intensity chromophore transitions, the chemical nature of which is very hard to define. All chromophores in bulk polymers are introduced either as impurities, including residual monomer, from thermal processes undergone in the preparation or work-up of the polymer or, most commonly, in the shape-giving processing of polymers. The best known photoactive species thermally produced in a polymer with an aliphatic backbone are carbonyl groups and particularly hydroperoxides. The latter groups are particularly important, as simple carbonyl groups are of little consequence in the photodegradation of purely aliphatic polymers such as polyolefins. The formation of these primary photoactive species in polymers is still a fertile ground for speculation and, for the purpose of this paper, we shall simply accept their presence in minute amounts as a given reality†.

A completely different behaviour, however, is expected from polymers which contain recurring units showing high absorption in the near u.v., such as polyphenylene oxides, the polyamides of aromatic diamines and polyvinyl carbazole or anthracene. The degradation of such polymers does not depend upon minute amounts of primary sensitizers and hence stabilization requires a different approach than that used in today's bulk polymers. However, in what follows, we will not deal with aromatic polymers of such high u.v.-absorption.

Let us now follow the course of events when a processed and shaped polymer is exposed to light.

The first step required in starting any photodegradation is of course the absorption of a photon by a primary sensitizer. The resulting excited singlet has the following possibilities: (i) it dissipates the accumulated energy by fluorescence and possibly internal conversion, (ii) it dissociates into hot fragments, (iii) it reacts with a partner to form hot reaction products, or (iv) it changes its multiplicity by intersystem crossing, whereby a large part of the energy absorbed still remains in the triplet formed. Of these possibilities the first is generally not harmful to the polymer and is therefore the preferred path of energy dissipation. Unfortunately, however, quantum yields in the second and third processes are, in most cases, small but significant. The further fate of the formed hot reaction products is important in

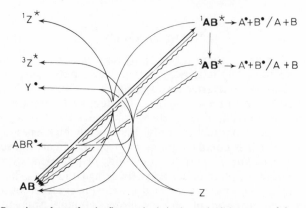

Figure 1. Reaction scheme for the first excited singlet and triplet states of the molecule AB.

---

† In the following sections we will call such chromophores originally present in polymers primary sensitizers. This is, however, without any implication to their actual mechanistic behaviour.

the photodegradation of polymers. If, upon dissociation of the excited singlet or by reaction with a partner, radicals are formed, normal autoxidation of aliphatic compounds—be they monomeric or polymeric in nature—ensues. All these reactions are so fast as to leave the excited singlet a mean lifetime of around 1 to 10 ns.

Intersystem crossing back to the ground state and phosphorescence of triplets is usually slow and gives a mean lifetime of around 10 to 1000 μs, i.e. triplets outlive singlets by a factor of 10^3 to 10^5. During their long life triplets have, of course, a good chance to enter chemical reactions. Again radical formation is particularly harmful, be it by dissociation (again the Norrish type I reaction in the case of excited carbonyl compounds) or by photoreductions involving a hydrogen-transfer. A schematic outline of these possibilities is given in *Figure 1*.

As pointed out before, autoxidation of the polymer induced by photolytically generated radicals is a major contributor to the observed overall degradation of the polymer. Any integral light protection of polymers must take this aspect into consideration. Accordingly, the following possibilities for light stabilization exist. They are listed in order of their action during the sequence of events in photodegradation: (i) u.v.-absorption, (ii) quenching of excited states, (iii) scavenging of photolytically produced radicals and (iv) prevention of radical formation by peroxide decomposition. The agents used to perform these actions are summarized in *Figure 2*.

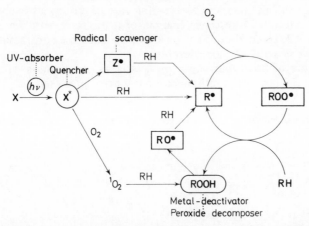

Figure 2. Schematic representation of the reactions involved in the light induced degradation of polymers with aliphatic backbones and name of the additives providing protection against the indicated reaction or chemical intermediate.

2. UV ABSORBERS

A large number of different chemical classes have been suggested as u.v.-screeners, mostly in the patent literature. However, only a few groups of compounds have found substantial use in industrial practice. They are the o-hydroxybenzophenones, 2-(2-hydroxyphenyl)-benzotriazoles and to a lesser extent salicylates and α-cyanocinnamic acid derivatives. Of the newer classes, the substituted oxanilides seem to be the most interesting ones[1]. However, their u.v.-absorption particularly in the long wavelength region is rather poor when compared to that of benzophenones and benzotriazoles. Their claimed effectiveness as light stabilizers seems therefore not entirely dependent upon u.v.-absorbing capacity. Due to the lack of thorough scientific investigation, detailed modes of action for such compounds are not yet clear.

A good u.v.-absorber must dissipate absorbed energy in a manner innocuous to the substrate. This means that dissipation of excitation energy should proceed at a faster rate than side reactions. Spectral data indicate that in 2-(2-hydroxyphenyl)-benzotriazoles and in o-hydroxy-benzophenones this energy dissipation process occurs solely in the singlet manifold. 2,4-Di-hydroxy-benzophenone shows neither phosphorescence nor triplet–triplet absorption[2]. The phosphorescence of 2-(2-hydroxy-5-methylphenyl)-benzo-triazole and of 2-hydroxy-5-methoxy-benzophenone is very weak, while 2-(2-hydroxy-3-tert.-butyl-5-methylphenyl)-5-chlor-benzotriazole does not phosphoresce at all[3]. Thus, intersystem crossing from the lowest excited singlet state S_1 to the lowest triplet state T_1 must be an unimportant de-activation process for S_1. Furthermore, the above mentioned compounds do not fluoresce at room temperature, while at 77 K only 2-(2-hydroxy-5-methylphenyl)-benzotriazole shows a very weak fluorescence[3]. Therefore, these molecules dissipate their excitation energy by a non-radiative singlet process, the rate of which must be significantly higher than natural fluores-cence decay and the intersystem crossing rate of the lowest excited singlet state. The natural fluorescence lifetime of a molecule with an extinction coefficient of 20 000 is about 5 to 10 ns, consequently the non-radiative decay rate must exceed 10^9 s^{-1} considerably.

The nature of this rapid decay process is, in our opinion, still open to discussion. So far, two mechanisms have been proposed to explain the given facts:

(A) The participation of an 'enol–keto-tautomerism' in the excited singlet state

'enol'-form   'keto'-form

It is assumed that, in the excited state, the 'keto'-form is more stable than the 'enol'-form, while in the ground state the 'enol' is more stable[4].

(B) The rotation of the hydroxyphenyl group. Thus the Franck–Condon factor between the excited singlet state and the ground state is enhanced (loose bolt effect) leading to a particularly high rate[5] for the—normally very slow—internal conversion from $S_1 \rightarrow S_0$.

The 'enol–keto-mechanism' is based on the observation of a very large Stokes shift of the fluorescence of o-hydroxyphenyl-pyrimidines[4]. Furthermore a number of examples are known in the literature which show that such proton-transfer reactions in the excited singlet state are indeed very rapid processes, with rates comparable or even much higher than the rate of the 'enol'-form fluorescence decay. Examples are derivatives of salicylic acid[6] or salicylidene anilines and 2-(2-hydroxyphenyl)-benzothiazole[7]. The 'keto'-fluorescence can be recognized by a red shift of about 4000 to 5000 cm^{-1} with respect to the 'enol'-fluorescence and in any case by a Stokes shift of about 10000 cm^{-1}. Our own investigations show that 'keto'-fluorescences are also formed in 2-(2-hydroxyphenyl)-benzoxazole and -imidazole. In polar solvents the imidazole derivative shows even in the ground state a small amount of 'keto'-form. In both compounds the ratio of the intensity of the 'enol' to 'keto'-fluorescence increases with increasing solvent polarity. For such equilibria which are shifted markedly by changing solvent polarity it has been established[8] that the enthalpy difference in the species taking part in such equilibria generally does not exceed 5 kcal/mol. Hence the energy difference between the excited 'enol'- and 'keto'-forms of these molecules also should not exceed this value of 5 kcal/mol. In fact Weller found[6] an enthalpy difference of 1 kcal/mol between the excited 'enol'- and 'keto'-forms of methylsalicylate.

One measure for this energy difference between the 'enol' and 'keto'-forms is certainly the difference between the acidity of the proton donor-part —the hydroxy group—and the basicity of the proton acceptor atom in these molecules. So the pK values for some of the interesting systems were determined. Direct measurements yielded values for the ground state while either the Förster-cycle method or fluorescence titration were used for the first excited singlet (pK*). The results in *Table 1* show that ground-state 'enol–keto'-equilibria can be observed if the difference between donor acidity and acceptor basicity is seven pH units or less. It is well known that in the first excited singlet state the acidity of phenolic hydroxyl groups is raised by about six units and the basicity of the carbonyl-oxygens or the ring-nitrogens functioning as proton acceptors is also raised by four to eight units. Therefore proton transfer in the excited state is much more probable than in the ground state.

The energy difference between the 'enol' and the 'keto'-forms in the ground state of methyl salicylate can be estimated from measurements by Weller[6] to be around 15 kcal/mol†. The difference in pK values between proton donor and acceptor in this molecule is estimated at 16 to 18 units[10]. This coincides very closely with the corresponding values for benzophenone and benzotriazole u.v.-absorbers (cf. *Table 1*). On the assumption that this equality in pK difference implies a similar difference in energy between the 'enol'- and 'keto'-forms in the ground state, i.e. 15 kcal/mol, an estimate of the energy difference of the excited 'enol' and 'keto'-forms can be made. Careful luminescence measurements reveal no 'keto'-fluorescence of the

---

† 15 kcal/mol $\simeq$ 80 kcal/mol ('enol'-fluorescence at 352 nm)
  $-$ 64 kcal/mol ('keto'-fluorescence at 443 nm)
  $-$ 1 kcal/mol (enthalpy difference between excited 'enol'- and 'keto'-forms)

benzophenones and benzotriazoles of *Table 1* up to 800 nm†. This means that if these *o*-hydroxybenzophenones and *o*-hydroxyphenylbenzotriazoles were to yield excited 'keto'-forms, their excitation energies would be at most 35 kcal/mol. This value is substantiated by the estimate that the rate of internal conversion ($S_1 \rightarrow S_0$) for rigid systems, which the 'keto'-forms are, will become comparable to the fluorescence decay only at higher wavelengths

Table 1. Properties of various u.v.-absorbing phenolic compounds.

The formulae of the compounds, the corresponding pK values in the ground state (pK), their difference (ΔpK) and the ones in the first excited singlet state (pK*) as well as the qualitatively spectroscopically determined presence of 'keto'-forms in the ground state (S_0 keto) and of 'keto'- and 'enol'-forms in the first excited singlet state (fluorescence enol, keto) are given in the columns from left to right.

	pK_1	pK_2	Δ pK	pK_1^*	pK_2^*	in ethanol		
						S_0 keto	Fluorescence enol	keto
2-(2-hydroxy-5-methylphenyl)-benzotriazole	−4.9	9.2	14	−0.5	(3)	—	+[1]	—
2-(4-hydroxyphenyl)-benzotriazole	−4.6	8.9		0	2	—	+	—
2-(2-hydroxyphenyl)-benzoxazole	0.9	9.7	9	4	4	—	+	+
2-(2-hydroxyphenyl)-benzimidazole	5.3	9.3	4	11	3	+	+	+
2-(2-hydroxyphenyl)-benzothiazole	<3[2]	≈8.5[2]				+[3]	+[3]	+[3]
2-hydroxy-4-methoxybenzophenone	−6.5	9.4	16	(1)	(3)	—	—	—
4-hydroxybenzophenone	−6.3	7.8		(−1)	(3)	—	—	—
2-methoxyphenol derivative	−6.5[2]	10[2]	16			—	+[4]	+[4]

[1] see text [2] ref. 10 [3] ref. 7 [4] ref. 6

† The low temperature fluorescence of 2-(2-hydroxy-5-methylphenyl)-benzotriazole at 396 nm and the very weak room-temperature fluorescence of 2-(2-hydroxyphenyl)-5-methoxy-6-methylbenzotriazole at 385 nm (in ethanol) are certainly 'enol'-fluorescences.

of the luminescence, i.e. at around 700 to 800 nm[9] which again corresponds to 35 to 41 kcal/mol. As the 'enol'-fluorescence of the hydroxybenzophenones and o-hydroxyphenylbenzotriazoles is around 400 nm or less ($\simeq$ 70 kcal/mol) the hypothetical excited 'keto'-form would have to be more stable than the excited 'enol'-form by at least 20 kcal/mol. This—when compared to the

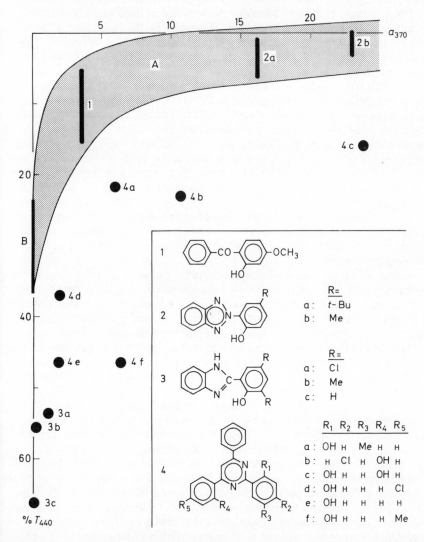

Figure 3. Action of various additives upon the light ageing of polyester resin. Plotted is the loss in transmission (at 440 nm) of 2 mm thick polyester plates versus the absorptivity of the u.v.-absorbing additive at 370 nm. (This latter value represents the empirically determined maximum of a wavelength dependent factor. This is the product of the sensitivity of polymer, its absorbance and output of the Fadeometer ® arc). The vertical bar (B) on the ordinate indicates the range of loss in transmission of additive free polyester plates. The shaded area (A) represents the loss area encountered with the best commercial and experimental u.v.-absorbers.

147

1 kcal/mol difference in the case of methyl salicylate—is a very unlikely value, particularly in view of the fact that the basicity of the proton acceptor in the benzophenones and benzotriazoles in question is very close to the one in methyl salicylate. From an energy point of view it is therefore very doubtful that 'keto'-forms exist during the deactivation process in these u.v.-absorbers and hence their contribution to energy dissipation seems highly questionable.

The existence of excited 'keto'-forms does not in itself make a compound a light-stabilizing u.v.-absorber. This is proved by the fact that o-hydroxy-phenyl-pyrimidines and o-hydroxyphenylbenzimidazoles have low light-stabilizing efficiency or even accelerate the discoloration of polyester resin as shown in *Figure 3*.

It is difficult to judge whether the rotation mechanisms can really explain the rapid dissipation of energy. The fact that 2-(2-hydroxy-5-methylphenyl)-benzotriazole does fluoresce weakly at 77 K but not at room-temperature indicates that the non-radiative deactivation process—whatever it is—must have an energy barrier; however, its activation energy is probably rather small. This finding is not in contradiction with a rotation which—in order to be operative—would require an activation energy of less than about 5 kcal/mol. Unfortunately no direct measurements are available today. Another indication in favour of the rotation mechanism is the fact that 2-hydroxy-4,6-di-tert.-butyl-benzophenone shows a strong phosphor-escence[11]. In this molecule in which rotation is strongly inhibited by steric hindrance, the dominant first excited singlet state deactivation process is obviously intersystem crossing to T_1.

While we favour the rotation mechanism for normal o-hydroxybenzo-phenones and 2-(2-hydroxyphenyl)-benzotriazoles, it should not be over-looked that other mechanisms can and must be operative in other systems. The high efficiency of 1-hydroxyxanthones as light stabilizers obviously requires a deactivation mode totally different from the loose bolt mechanism.

The predominance of a rapid and harmless deactivation process for the first excited singlet state—which process in our experience should be non-radiative—is, however, not the only pre-requisite of a technically useful u.v.-absorber. In addition to the spectral characteristics which have been discussed elsewhere[12], a truly monomolecular dispersion of the screener in the polymeric substrate is necessary. If this pre-requisite is not fulfilled, the activity of a specific compound is lower than expected from its behaviour in solution. This fact allows easy determination of u.v.-absorber 'functional compatibility', in contrast to the commonly used 'visual compatibility', i.e. the lack of an observable formation of an additional stabilizer phase in the polymer. In order to assess 'functional compatibility' the experimentally determined absorbance of a film or plaque containing u.v.-absorber is compared to the one calculated from the spectral data of the u.v.-absorber in a solvent with characteristics comparable to the poly-meric substrate. In this general procedure the use of a series of homologous compounds—all containing the same chromophore—is particularly recom-mendable. *Figure 4* presents data obtained with o-hydroxyphenyl-benzo-triazoles into which alkyl sidechains of varying length and/or branching were introduced by means of an ester group. The absorbance of approxi-mately 0.1 mm thick low density polyethylene films containing 0.2 per cent

Compatibility in polyethylene

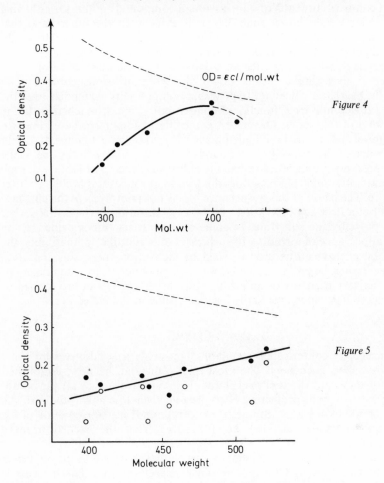

$$OD = \varepsilon c l / mol.wt$$

Figure 4

Figure 5

Figures 4 and 5. Absorbance of polyethylene films versus the molecular weight of u.v.-absorber. Formulae of absorbers:

Figure	R_1	R_2
4	$-CH_2CH_2COOR$	$-H$
5	$-CH_3$	$-CH_2NHCOR$

(R standing for alkyl rests)

Concentration of u.v.-absorber 0.2 per cent. The dashed line represents the calculated absorbance. Further explanations appear in the text.

149

u.v.-absorber is plotted versus the molecular weight of the absorber used. 'Visual compatibility' was obtained with all compounds of this series having a molecular weight larger than 300. It is evident on the other hand that 'functional compatibility' is restricted to compounds of a molecular weight 400 ± 20 corresponding to octyl esters. For such compounds the observed absorbance equals the expected absorbance within the limits of error.

Figure 5 gives similar results for *o*-hydroxybenzotriazoles containing an amide linkage in a sidechain (●). 'Visual compatibility' extended over the whole range while true 'functional compatibility' is not reached with any compound of this series. Of interest is the effect of light on these films. An exposure of 100 hours in a Fadeometer® (○) produces a significant loss in absorbance of these specific compounds in low density polyethylene, while the same exposure in other polymeric substrates, such as cellulose acetate films and polystyrene plaques, does not lead to any measurable changes after exposure. The cause of this absorbance loss is not photolysis of the chromophores but a slow agglomeration or even crystallization of the u.v.-absorber in the polyethylene substrate. In some extreme cases this is evidenced by visual appearance of turbidity. In such cases of borderline compatibility, the excitation energy can be used to yield the activation energy of separation of these rather larger molecules from their submicroscopic agglomerates or the activation energy of diffusion. Thus larger aggregates are formed or even crystallites which eventually become visible to the naked eye.

3. QUENCHERS

A number of commercially available light-protecting additives for polymers are called quenchers. The most important and, by now, established group of these are the nickel chelates. The main feature of all these substances is their light-protecting effect despite their low absorptivity in the region of 300 to 400 nm†. But are these compounds in the true sense of the word quenchers, i.e. do they accept energy from the excited primary sensitizers?

Phenomenologically two different kinds of quenching can be distinguished in photochemistry. (i) One is long-range energy-transfer. This process is normally observed in the quenching of excited singlet states. It is found to operate only when distances between sensitizer and quencher (R_O)‡ are 50 Å or greater. (ii) The other types are contact transfers. These are mechanisms of a different nature, but all of which are effective when the distance between quencher and sensitizer is 15 Å or less. The quenching process is successful only if the quencher is or gets within quenching distance of the excited sensitizer within the latter's lifetime. High diffusion constants in a substrate, i.e. good mobility of quencher and quenchee, and long lifetime of the excited sensitizer may therefore enlarge the apparent action sphere of

---

† Obviously the normal u.v.-absorbers can act as potential quenchers, but since their activity depends on a high molar extinction coefficient in the near u.v., they are normally not termed quenchers.

‡ In the following R_O stands for the mean distance between sensitizer and quencher at the moment of the energy transfer act. R_{SQ}, however, is meant to indicate the distance between sensitizer and quencher, calculated from concentration.

the quencher or in other words lower the concentration of quencher necessary to observe a certain effect. Obviously this can be accurately calculated for each specific case. In order to get an overall impression of the general situation *Figure 6* is presented. The assumptions underlying this

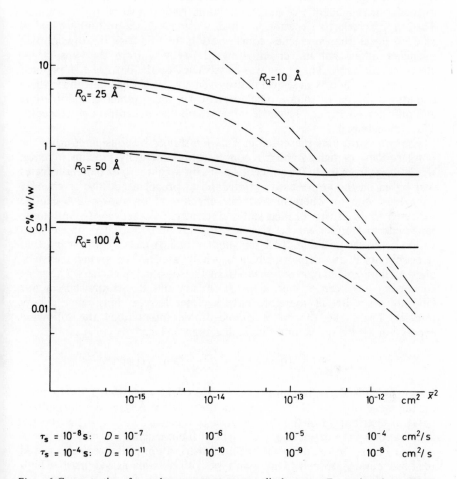

Figure 6. Concentration of quencher versus mean square displacement. For explanation, see text.

figure are (i) a freely diffusing quencher of molecular weight 500 and (ii) a fixed sensitizer with an active site diameter of 5 Å†.

In this graph, the quencher concentration is plotted as a function of the mean square displacement ($\bar{x}^2$), i.e. the concentration needed to allow quenching within the lifetime τ_s in a substrate characterized by a diffusion

---

† Such fixed sensitizers would correspond to sensitizing moieties attached to the polymer chains in polymeric substrates.

constant D. Each curve represents a specific distance (R_Q) at which the energy transfer becomes operative.

The dashed lines give the best possible case (unit probability), i.e. the biggest enlargement of quencher action sphere, which results if each quencher is surrounded by sensitizers. Under these circumstances any movement of the quencher, independently of direction, would lead to a successful encounter. The full lines reflect a situation in which sensitizer concentration is equal to or even lower than quencher concentration. In this case the direction of quencher diffusional movement—toward or away from the sensitizer— starts to play a role. The figure was constructed on the basis of a 20 per cent probability of success in diffusional movement. It is seen that under these conditions the mean square displacement, i.e. the product of half-lifetime and diffusion coefficient, has little influence on the concentrations needed to effect quenching.

The concentrations presented in *Figure 6* should be considered from the point of view of light stabilizer concentrations actually used in practice, which range from 0.1 to 0.5 per cent weight by weight. *Figure 6* demonstrates that based on the above assumptions and at practical additive levels only quenchers with an operational mode effective at or above 50 Å can be expected to deactivate excited states efficiently. In other words, only long-range energy transfer can be expected to contribute to excited-state deactivation with the usual half-life of singlets and triplets. This means that quenching of excited triplets, which is usually ascribed to contact transfers, plays a minor role in light stabilization by the so-called quenchers. The same conclusion has been arrived at by B. Felder and R. Schumacher in our laboratories[13]. Let us therefore have a closer look at long-range energy transfer. From the theory of dipole–dipole interactions the following expression for the quenching rate k_Q has been deduced[14]

$$k_Q = 3.7 \times 10^{22} \times (\Phi_S/\tau_S R_Q^6) \int\limits_0^\infty f_S(v)\, \varepsilon_Q(v)\, dv/v^4$$

In this formula Φ_S is the fluorescence yield of the sensitizer, τ_S its lifetime, R_Q the mean distance of sensitizer and quencher in Å, f_S is the normalized spectral distribution of the fluorescence of the sensitizer, ε_Q is the spectral distribution of the extinction coefficient of the quencher and v is the wave-number in cm^{-1}. Assuming the spectral distribution of fluorescence f_S and the absorption ε_Q to be of Gaussian type with maxima at 345 nm and half-width of $4000\ cm^{-1}$† and considering that the ratio of Φ_S/τ_S is equal to the fluorescence decay rate k_f, one finds the following proportionality between the rate of quenching and the rate of fluorescence decay

$$k_Q = 3.7 \times 10^4 \times \varepsilon_{Qmax} \times k_f/R_Q^6$$

A quencher concentration of 0.02 mol/l., say, one per cent at a molecular weight of 500, corresponds to $R_{SQ} \approx 40$ Å. Experience shows that the quench-radii calculated by the theoretical model are about a factor of two

---

† These are reasonable assumptions for aliphatic ketones as sensitizers and nickel chelates or u.v.-absorbers as energy acceptors.

lower than those determined from the concentration. Thus

$$k_Q = 5 \times 10^{-4} \times \varepsilon_{Qmax} \times k_f$$

This expression leads to the conclusion that additives with extinction coefficients of 2 000 and higher can quench excited singlet states with rates equal to or higher than the fluorescence decay rate†. However, the very low fluorescence yields of primary sensitizers, such as ketones and peroxides (e.g. diethylketone $\Phi_f = 0.01^{15}$), mean that non-radiative deactivation processes of the excited singlet state are faster and more important than the fluorescence. Hence k_Q must be much larger than k_f if quenching is to compete with these non-radiative deactivation processes of the sensitizer. Thus a compound acting as a long-range quencher must have an ε_{Qmax} in excess of 10 000 around 340 to 350 nm. Such compounds, however, are the typical u.v.-absorbers known to the trade.

To conclude this section, we would like to stress that at the additive levels usually employed in polymeric substrates, the quenching of excited triplet states by contact transfer does not appear to be a major factor in the light stabilization of polymers with aliphatic backbones. Quenching of excited singlet states by dipole–dipole interaction, however, can be a major factor in the stabilization process. An entirely different situation may exist in polymers with strongly luminescing moieties in the backbone or as pendant sidechains.

4. AMINE STABILIZERS

In the scientific literature as well as in patents, specifically substituted derivatives of heterocyclic amines have been suggested as polymer additives. Early publications have concentrated on N-oxyl free radicals such as:

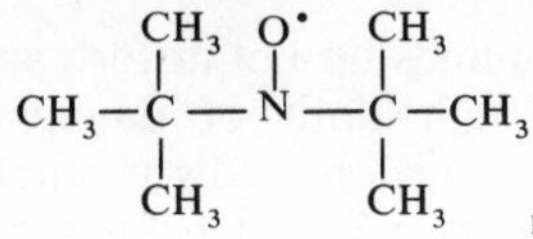

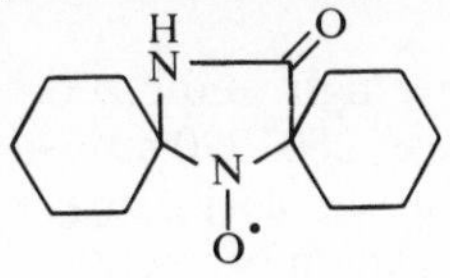

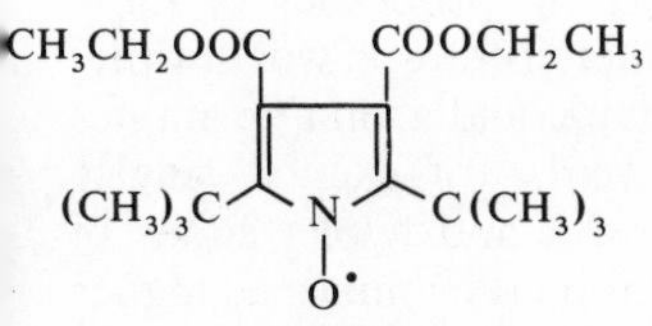

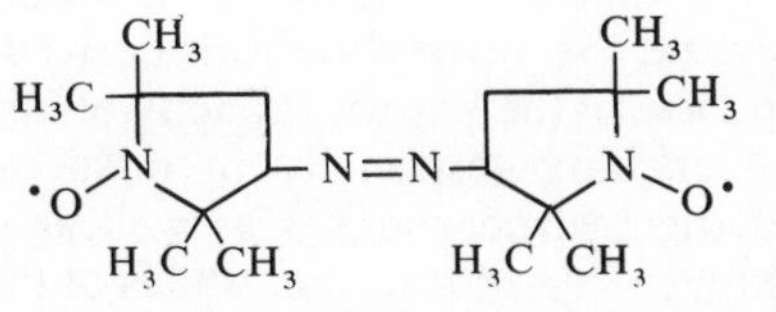

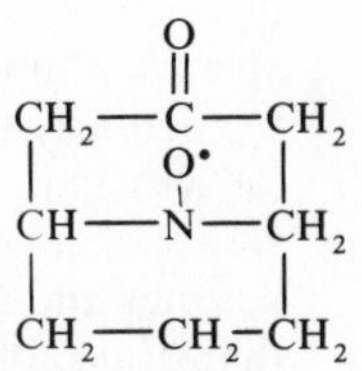

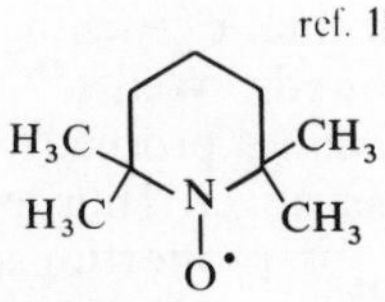

† The work of Chien and Conner[15] provides evidence that the singlet state of diethylketone can be quenched by the nickel chelate of 2,2′-thiobis-[4(1,1,3,3-tetramethylbutyl)-phenol], which has an ε_Q (313 nm) of 4000. However, the concentration (0.5 mol/l.) of diethylketone (mol. wt 128.22) used is so high as to prove little for practical conditions in actual polymers.

Excellent review articles on nitroxyls have been published by K. Murayama[20] and E. G. Rosantzev et al.[21]. The latter has contributed significantly to the general knowledge of the nitroxyl free radicals. These compounds are very useful spin probes for polymers[21]. The application of nitroxyl radicals in the stabilization of polymers was originally considered on the basis of their capability to trap the free radicals essential in the degradation of polymers[22]. Accordingly the simple nitroxyls were thought to be good antioxidants. However, their efficiency in preventing thermal oxidative degradation is not sufficient to allow commercial usage for this purpose. Later on the usefulness of these additives in polymer light stabilization, particularly polyolefins, was discovered[23]. Their efficiency in suppressing photodegradation of polyolefins can—depending upon the specific substitution—be quite remarkable. In this respect our own experiments confirm[24] statements made in and the claims of various patents. However, the colour of nitroxyl radicals —yellow to red—effectively prevents usage as commercial plastic additives in the concentration range normally utilized, e.g. around 0.5 per cent based on the polymer.

A big step forward was the surprising finding by chemists of Sankyo Company Limited that not only the free radicals described, but also specific free amines, e.g.

are effective light stabilizers[25]. As these compounds do not absorb appreciably above 280–290 nm—the short wavelength limit of daylight—the question arises as to how such compounds function as light stabilizers. In spite of the aforementioned conclusions concerning quenching in polymers, it was felt that investigation of the quenching properties of these specific amines and the nitroxyl radicals derived therefrom was worthwhile; particularly in view of the known capacity of aliphatic and aromatic amines for efficient quenching—especially in polar solvents—of excited singlet states of aromatic hydrocarbons[26] as well as singlet and triplet states of ketones and oxygen[27]. As the excited states of the aliphatic amines lie higher than those of the sensitizer, normal quenching mechanisms cannot explain the efficiency of these compounds. Weller[26] showed that the singlet deactivation mechanism of the amines proceeds by an excited charge-transfer complex between amine and sensitizer. Thus for a given sensitizer in a given solvent the quenching rate k_Q is proportional to the ionization potential of the amine. In order to determine whether tetramethyl-piperidine derivatives of the patent literature could deactivate singlets in polymers and what factors affect their quenching efficiency, ionization potentials and quenching constants for a series of amines were measured. The ionization potentials were determined by ground state charge-transfer complexes with iodine as an electron acceptor. The quenching constants presently reported are the gradients $k_Q \tau_s$ of Stern–Vollmer plots using fluorenone as a sensitizer in acetonitrile.

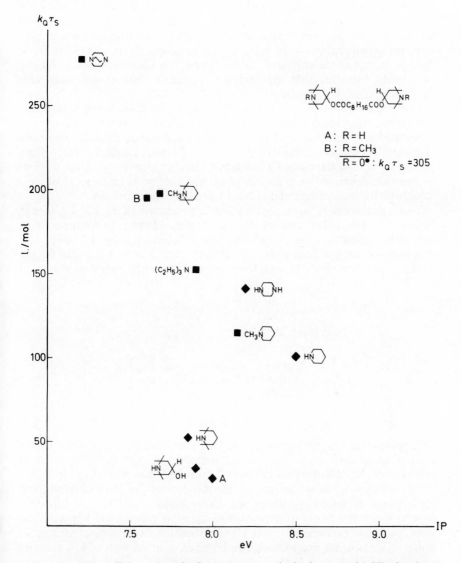

Figure 7. Quenching efficiency ($k_Q \tau_s$) for fluorenone versus ionization potential (IP) of various amines. For explanation, see text.

Figure 7 shows that 2,2,6,6-tetramethylpiperidines—independent of the substituent in position 4—have ionization potentials similar to triethylamine, namely 7.85 to 8.0 eV. The corresponding *N*-methyl derivatives have lower ionization potentials by about 0.2 eV or 5 kcal/mol. The nitroxyl radicals did not form measurable amounts of iodine complexes; consequently their ionization potential could not be determined. *Figure 7* shows, furthermore, that the ionization potential is not the only factor affecting the quenching constant. Apparently the steric hindrance of the lone pair on the nitrogen by the α-position substituents lowers the probability of a successful sensitizer quencher encounter. N-Methylation increases the quenching constants remarkably (4- to 6-fold). The reason for this enhancement in spite of further steric hindrance of the nitrogen is not clear at the moment. One could speculate that the exciplex encounter distance increases with decreasing ionization potential.

The reciprocal of the quenching constant gives the quencher concentration needed to deactivate one half of the excited sensitizers before they fluoresce. In apolar solvents the quenching constants are about 50 times smaller than in acetonitrile (heptane : triethylamine $k_Q\tau_s = 3$ l./mol, 1,2,2,6,6-pentamethyl-piperidine $k_Q\tau_s \approx 1$ l./mol). Thus it is not possible for even the most efficient of these amines to act as quencher of excited singlets of primary sensitizers in polyolefins, considering the normal stabilizer concentration.

This conclusion is borne out by the experiments of H. Lind[28] in our laboratory. The initial rate of disappearance of 1-phenyl-decanon-(1) as well as the formation of the photolysis products acetophenone and octene-(1)† in n-heptane using a high pressure mercury lamp was not changed by the addition of compounds Ia and Ib in the concentration range of 4×10^{-3} to 1.3×10^{-1}.

$$H \quad OCO(CH_2)_8COO \quad H$$

CH₃ ... CH₃ CH₃ ... CH₃

CH₃ N CH₃   CH₃ N CH₃

R   R

I
a: R = H
b: R = CH₃
c: R = O

This proves that the free amines Ia and Ib, under the experimental conditions used, have no quenching ability on the first excited singlet state of carbonyl compounds. In addition they cannot quench the triplet states.

The nitroxyl Ic, however, completely suppresses the Norrish photolysis of 1-phenyl-decanone-(1) under the same conditions. This means that nitroxyls can effectively quench excited states of araliphatic ketones. The most probable process is triplet state quenching as singlet quenching in polar solvents is not great enough to be efficient in solvents of low polarity (cf. *Figure 7*, Ic: $k_Q\tau_s = 305$).

From these experiments it can be concluded that in the photodegradation of polymers such as polyolefins the quenching of excited carbonyls as primary sensitizers is not a very important process. This is further supported

---

† The reaction in degassed solution was followed by VPC up to ten per cent conversion of the starting ketone which had an initial concentration of 2.5×10^{-3}.

by the fact that the addition of low volatility aliphatic ketones† does not significantly change the photodegradation rates of polypropylene stabilized only with the amount of antioxidant necessary to ensure reproducibility of the films pressed[28].

What other possibilities exist then which contribute to the light induced breakdown of polymers with aliphatic backbones? One agent which has been mentioned frequently in the last few years is singlet oxygen. Consequently Felder and Schumacher[13] and Belluš, Lind and Wyatt[29] have investigated the ability of various types of compounds to quench singlet oxygen, produced either photochemically, i.e. by rose bengal and methylene blue, or chemically, i.e. by the hydrogen peroxide/hypochlorite reaction. For this purpose the disappearance of various singlet oxygen scavengers such as rubrene was followed.

Ia as well as some other secondary amines IIa, b and c did not show

a: R_1 = H R_2 = H
b: R_1 = H R_2 = OH
c: R_1 = OH R_2 = OPO(OEt)$_2$

significant quenching of singlet oxygen while the corresponding nitroxyls Ic as well as III a, b, c had a marked effect on the deactivation of the photo-induced oxygen addition of the indicators used. The N-methyl compounds IIb and IVa, b, c also slowed down the disappearance rate of the singlet oxygen indicators; however, it could be shown that oxidative demethylation of the tertiary amines by singlet oxygen occurred under these conditions[29]. This renders the detection and quantification of any quenching effect impossible.

In *Figure 8* the ratio of rubrene concentration in samples with and without additive after irradiation for ten minutes is given for various classes of compounds. As indicated before, the apparent high efficiency of N-methyl-2,2,6,6-tetramethylpiperidines is at least partly, if not wholly, due to chemical consumption of singlet oxygen. The next most effective class are the nickel chelates followed by the tetramethylpiperidine-N-oxyls. The secondary amines of the tetramethylpiperidine series as well as antioxidants of the sterically hindered phenol class have little or no quenching effect at the additive concentration (5×10^{-2} mol/l. in ethanol/benzene 1:1) used.

---

† 0.2 per cent each of nonadecanone-(2), nonadecanone-(9) and stearone corresponding to a carbonyl value (absorbance at $1\,718$ cm^{-1}) in 0.1 mm films of 0.035, 0.035 and 0.02 respectively. A control film of the same composition and thickness but without added ketones shows a complete loss of mechanical properties at carbonyl values of 0.1 to 0.2.

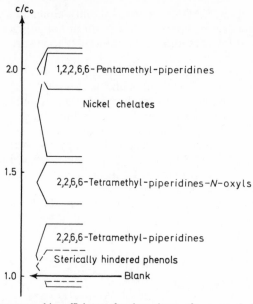

Figure 8. Singlet oxygen quenching efficiency of various classes of compounds. Further explanation is given in the text. The figure is constructed with results from the following compounds (from ref. 13):

(i) 2,2,6,6-Tetramethyl-piperidines:

R_1:H R_2:H   (IIa)
    H   OH   (IIb)
    H $OCOC_{18}H_{37}$ (IId)
    H $OCOC_4H_8)_2$ (Ia)
R_1:Me R_2:$OCOC_{18}H_{37}$ (IVd)

R_1:O· R_2:OH   (IIIb)
        R_2:$OCOC_4H_8)_2$ (Ic)

(ii) Nickel chelates

$$\left[HO-\underset{\times}{\underset{\times}{\bigcirc}}-CH_2PO_2OC_2H_5 \right]_2 Ni$$

(iii) Sterically hindered phenols

$$R_3 = -\underset{\times}{\underset{\times}{\bigcirc}}-OH$$

$$[R_3-(CH_2)_2COOCH_2]_4C \qquad R_3-(CH_2)_2COOC_{18}H_{37}$$

$$R_3-COOC_{18}H_{37} \qquad R_3-CH_2-PO(OC_2H_5)_2$$

This allows the following conclusions:

(a) The quenching of primary sensitizers is not necessarily very important in antioxidant containing substrates, as has been pointed out above in the case of carbonyls.

(b) The deleterious effect of singlet oxygen is either not very important or its consequences can be healed in subsequent stabilization steps.

(c) The secondary and tertiary amines act by mechanisms of stabilization other than u.v.-absorption and quenching.

EXPERIMENTAL SECTION

The ground state pK-values were determined spectroscopically in buffer solutions or sulphuric acid with 20 per cent ethanol. The H_0-values are based on the acidity function of Dolman and Stewart[30] (calibrated with diphenyl amines). The gradients of the logarithmic titration curves were 0.85 to 0.95 for benzotriazoles and 1.02 to 1.08 for benzophenones.

The pK*-values were calculated by the 'Förster cycle'[31] with the O—O transitions as averages of the maxima of the fluorescence and the absorption bands. For 2-(4-hydroxyphenyl)-benzotriazole and for 2-(2-hydroxyphenyl)-benzoxazole this value corresponded within ± 0.5 unit with the one from fluorescence titration. For the non-fluorescent benzophenones the pK* had to be determined from the absorption spectra of acid and base alone. These values are inaccurate and are therefore given in brackets in *Table 1*.

The ionization potentials were calculated from the absorption maxima of the charge-transfer-spectra of the amines with iodine as the electron acceptor in heptane. These relative values were calibrated with absolute data obtained by photoelectron spectroscopy in the gas phase[32] (four calibration points, adiabatic ionization potentials). The quenching constants were obtained from the gradients of 'Stern–Vollmer plots' with fluorenone as sensitizer in acetonitrile:

$$\Phi_0/\Phi_Q = 1 + k_Q \tau_s [Q]$$

Φ_0 is the fluorescence intensity of the sensitizer without quencher and τ_s the lifetime of the first excited singlet state of the sensitizer. Φ_Q is the fluorescence intensity in the presence of a quencher in the concentration $[Q]$. k_Q is the quenching rate constant. In heptane, only triethylamine and 1,2,2,6,6-pentamethyl-piperidine were soluble enough to produce a measurable decrease of the fluorescence of fluorenone.

REFERENCES

[1] FR patent 1516276 (priority date 7.2.66) CIBA;
DAS 1907403 (priority date 19.2.68) Sandoz.
[2] A. Beckett and G. Porter, *Trans. Faraday Soc.* **59**, 2051 (1963).
[3] F. G. Kirkbright, R. Narayanaswamy and T. S. West, *Anal. Chim. Acta*, **52**, 237 (1970).
[4] J. E. A. Otterstedt and R. Pater, *J. Heterocyclic Chem.* **9**, 225 (1972) and references cited therein.
[5] H. J. Heller, *Europ. Polymer J.—Supplement*, 122 (1969).
[6] H. Beens, K. H. Grellmann, M. Gurr and A. Weller, *Disc. Faraday Soc.* **39**, 183 (1965).
[7] M. D. Cohen and S. Flavian, *J. Chem. Soc. B*, 317 and 321 (1967); for 2-(2-hydroxyphenyl)-benzothiazole see also
D. L. Williams and A. Heller, *J. Phys. Chem.* **74**, 4473 (1970).
[8] R. Potashnik and M. Ottolenghi, *J. Chem. Phys.* **51**, 3671 (1969).
[9] W. Siebrandt, *J. Chem. Phys.* **49**, 1860 (1968).
[10] 'Stability Constants', *Spec. Publ. Nos 17* and *25* of the Chemical Society, London;
G. Kortüm, W. Vogel and K. Andrussow, *Dissoziationskonstanten organischer Säuren*, Butterworths: London (1961);
K. Schwarzenbach, CIBA–GEIGY AG, private communication.
[11] E. O'Connelly Jr, *J. Amer. Chem. Soc.* **90**, 6550 (1968).
[12] H. J. Heller and H. R. Blattmann, *Pure Appl. Chem.* **30**, 145 (1972).
[13] B. Felder and R. Schumacher, *Angew. Makromol. Chemie*, submitted for publication.
[14] T. Förster, *Disc. Faraday Soc.* **27**, 7 (1959).
[15] J. C. W. Chien and W. P. Conner, *J. Amer. Chem. Soc.* **90**, 1001 (1968).
[16] US Pat. No. 3163677 (priority date 7.8.1961) American Cyanamid Comp.
[17] Brit. Pat. No. 1194402 (priority date 15.6.1966) Sankyo Comp. Ltd.
[18] Brit. Pat. No. 1118160 (priority date 26.11.1964) ⎫ Commissariat
US Pat. No. 3494930 (priority date 27.6.67) ⎬ à l'Energie
Fr. Pat. No. 1579553 (priority date 4.4.68) ⎭ Atomique (France).
[19] US Pat. No. 3334103 (priority date 13.6.1962) American Cyanamid Comp. and G. I. Likhtenshtein, *Zh. Fiz. Khim.* **36**, 750 (1962).
[20] K. Murayama, *J. Synth. Chem. Japan*, **29**, No. 4 (1971).
[21] E. G. Rosantsev and V. D. Sholle, *Synthesis*, 190 and 401 (1971) (references for spinprobes see p 414).
[22] M. B. Nieman, *Aging and Stabilization of Polymers*, p. 33; Consultants Bureau, New York (1965).
[23] Brit. Pat. No. 1130799 (priority date 26.11.1965);
Brit. Pat. No. 1130386 (priority date 26.11.1965);
Sankyo Comp. Ltd. and Asahi Chem. Comp. Ltd.
[24] Private communication of H. Müller, CIBA–GEIGY AG.
[25] Brit. Pat. No. 1196224 (priority date 22.6.1967) Sankyo Comp. Ltd.
[26] A. Weller and D. Rehm, *Ber. Bunsengesellschaft*, **73**, 834 (1969) and references cited therein.
[27] L. A. Singer, *Tetrahedron Letters*, 923 (1969).
G. A. Davies, P. A. Carapellucci, K. Szoz and J. D. Gesser, *J. Amer. Chem. Soc.* **91**, 2264 (1969);

E. A. Ogryzlo and C. W. Tang, *J. Amer. Chem. Soc.* **92**, 5034 (1970).
[28] H. Lind, CIBA–GEIGY AG, private communication.
[29] D. Bellus, H. Lind and J. F. Wyatt, *Chem. Commun.,* submitted for publication.
[30] D. Dolman and R. Stewart, *Canad. J. Chem.* **45**, 904 (1967).
[31] A. Weller in *Progress in Reaction Kinetics* (G. Porter, ed.) Vol. I, p 187; Pergamon Press: London (1961).
[32] W. J. Wedenejew, L. W. Gurwitsch, W. H. Kondratjew, W. A. Medwedew and E. L. Franke-witsch, *Energien chemischer Bindungen, Ionisationspotentiale und Elektronenaffinitäten.* VEB Verlag für Grundstoffindustrie: Leipzig (1971).

161

MECHANISMS OF ANTIOXIDANT ACTION.
PHOSPHITE ESTERS

K. J. Humphris and G. Scott

The University of Aston in Birmingham, Gosta Green, Birmingham, UK

ABSTRACT

Catechol phosphite esters are powerful stabilizers for polymers and appear to behave differently from simple alkyl or aryl phosphites in that they destroy hydroperoxides in a Lewis acid catalysed reaction. The effective catalyst is formed from the starting ester by reaction with hydroperoxides in a series of chemical reactions which during the initial stages involve the formation of free radicals. Some of the isolated reaction products do not show this behaviour and are themselves powerful thermal and u.v. stabilizer antioxidants.

Esters of phosphorous acid have become increasingly important in recent years as antioxidants and stabilizers. A long established use has been in the stabilization of uncured synthetic rubbers and tris-nonylphenylphosphite I(a) is still one of the favoured cheap gel inhibitors for styrene, butadiene rubber[1].

$$P(OR)_3$$

I

(a) $R = $ —⟨benzene ring⟩—C_9H_{19}

(b) $R = $ alkyl

More recently, the alkyl phosphites I(b) and the mixed alkyl–aryl phosphites have assumed importance in polyolefins[2] and polyvinylchloride[3] as u.v. stabilizers. An important subgroup within the above general class of phosphite esters which has assumed particular importance in polyolefins is the catechol phosphite esters (II), which are particularly effective when R is hindered phenyl.

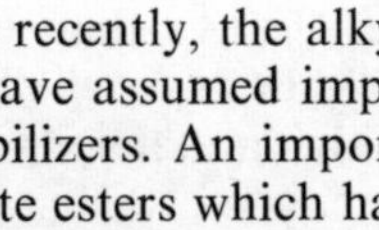

II

(a) $R = -CH(CH_3)_2$

(b) $R = $ ⟨cyclohexyl ring⟩

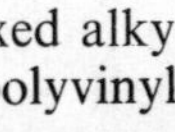

(c) $R = $ ⟨phenyl ring with t-Bu, t-Bu, Me substituents⟩

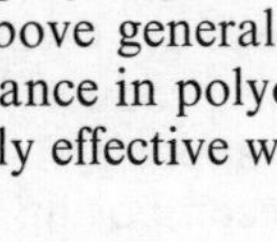

Extensive technological investigations have been carried out by Kirpichnikov and his co-workers, particularly in polyolefins[4–7], and in rubbers[8, 9] on the relationship between structure and activity. These workers have shown[7] that the hindered phenyl phosphite II, R = 2,4,6-tri-tert-butylphenyl was the most effective structure examined and that antioxidant effectiveness paralleled reactivity toward the 'stable' free radical diphenylpicryl hydrazyl. The high antioxidant efficiency of the catechol hindered-phenyl phosphites has been confirmed in the present studies (see later).

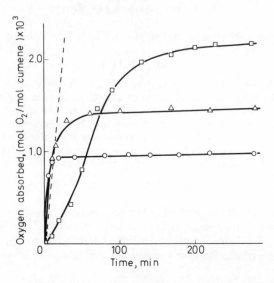

Figure 1. Inhibition of AZBN-initiated oxidation of cumene in oxygen at 50°C by 0.2M catechol phosphites. $[AZBN]_0 = 0.02M$; □ hindered phenyl phosphite; △ isopropyl phosphite; ○ phenyl phosphite – – – – no phosphite.

Rysavy and his co-workers[10, 11] have shown that in polypropylene, the aryl catechol phosphites (II, R = aryl) are more effective than the alkyl analogues (II, R = alkyl); however, there is some doubt as to whether these compounds are themselves responsible for the antioxidant activity or some hydrolysis product[12, 13].

There appears to be no agreement as to whether these phosphite esters act by decomposing hydroperoxides or by removing chain propagating free radicals from the autoxidizing medium. Previous studies have shown[14] that a useful and sometimes diagnostic method of distinguishing between the two mechanisms is to compare the kinetics of hydrocarbon oxidation in the presence of an azo initiator and a hydroperoxide initiator. In general antioxidants which function exclusively as hydroperoxide decomposers are inactive in the former system.

In cumene initiated by azo-bis-iso-butyronitrile (AZBN) it was found that the phenyl ester II(b) rapidly reduced the rate of oxidation to zero after an initial pro-oxidant period (*Figure 1*). The iso-propyl ester II(a) was less effective and the hindered phenyl ester, II(c), required a large oxygen absorption before it became an effective antioxidant. By contrast, triphenylphosphite

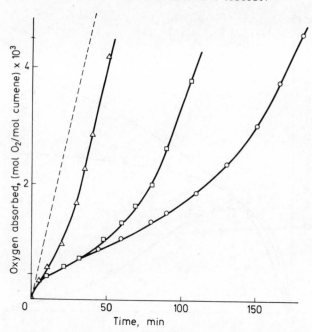

Figure 2. Inhibition of AZBN-initiated oxidation of cumene in oxygen at 75°C by triphenyl phosphite. $[AZBN]_0 = 0.01\text{M}$: $\triangle 0.022\text{M}$; $\square 0.035\text{M}$; $\bigcirc 0.051\text{M}$; ———— no phosphite.

under the same conditions was auto-accelerating after a slight induction period (*Figure 2*).

With cumene hydroperoxide as initiator, more pronounced pro-oxidant effects were observed with the catechol hindered-phenyl phosphite (see *Figure 3*) which was selected for more detailed study because of its hydrolytic stability (see *Figure 4*). In addition gas evolution was observed with higher concentrations (*Figure 3*) and this behaviour was lacking in the case of triphenyl phosphite (*Figure 5*) which rapidly reduced the rate of oxidation to a constant value after an initial pro-oxidant effect. The gas evolution was shown to be associated with the reaction of hydroperoxide with phosphite since it also occurred in chlorobenzene which is inert to oxidation (see *Figure 6*). Gas evolution also occurred in the absence of oxygen (see *Figure 7*), and the fact that oxygen is absorbed by the oxidizable substrate (tetralin) indicates that free radicals are involved in its formation.

Comparison of the behaviour of the catechol phosphite esters with triphenyl phosphite led to the conclusion that whereas the latter was acting by stoichiometric destruction of hydroperoxide

$$(\text{PhO})_3\text{P} + \text{ROOH} \rightarrow (\text{PhO})_3\text{P} = \text{O} + \text{ROH}$$

the former, or products derived from them, appeared to destroy hydroperoxides in a catalytic manner. This was unexpected since only the stoichiometric reaction of hydroperoxides with phosphite esters has previously been

165

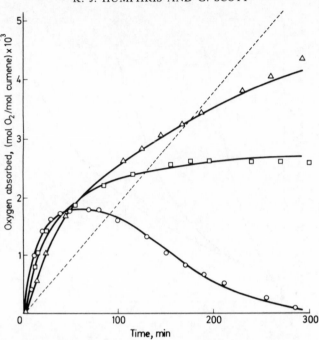

Figure 3. Inhibition of cumene hydroperoxide-initiated oxidation of cumene in oxygen at 75°C by catechol hindered-phenyl phosphite. $[CHP]_0 = 0.2M$; $\triangle$ $2 \times 10^{-3}M$; $\square$ $1 \times 10^{-2}M$; $\bigcirc$ $2 \times 10^{-2}M$; ---- no phosphite.

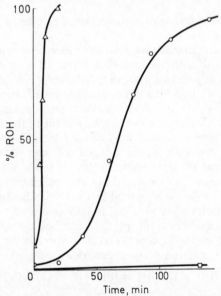

Figure 4. Hydrolysis of catechol phosphites (0.02M) in 95% (v/v) aqueous ethanol at 75°C. $\square$ catechol hindered-phenyl phosphite; $\triangle$ catechol phenyl phosphite; $\bigcirc$ catechol isopropyl phosphite.

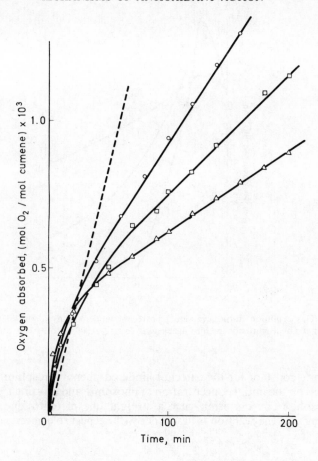

Figure 5. Triphenyl phosphite-inhibited oxidation of cumene initiated by 0.2M CHP in oxygen at 75°C. ○ 0.01 M; □ 0.02 M; △ 0.03M phosphite; – – – – no phosphite.

reported[15]. Detailed studies have confirmed this. *Figure 8* shows the effect of the three catechol phosphite esters in the decomposition of cumene hydroperoxide. All are catalysts for hydroperoxide decomposition and show good first order kinetics although the hindered-phenyl phosphite, II(c) appears to exhibit more complex behaviour initially. The apparent first order rate constants are listed in *Table 1*.

Table 1. Apparent first order rate constants for the decomposition of 0.2M cumene hydroperoxide in nitrogen at 75° by 0.02M catechol phosphites (II)

Phosphite	II (a)	II(b)	II(c)
k, S^{-1}	8.0×10^{-4}	2.5×10^{-3}	6.7×10^{-5}

167

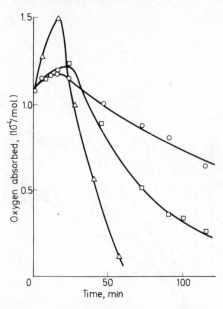

Figure 6. Gas evolution during reaction of catechol hindered-phenyl phosphite with 0.2M cumene hydroperoxide in chlorobenzene at 75°C. ○ 0.0095M; □ 0·021M; △ 0.04M.

The rate constant for the catechol hindered-phenyl phosphite was found to depend on the initial concentration of phosphite and plotting $\log k$ against log phosphite concentration gave a straight line of unit slope (*Figure 9*) indicating that the reaction is first order with respect to a species (P) formed

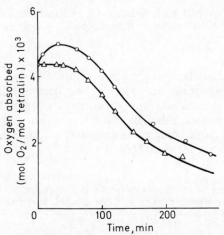

Figure 7. Gas evolution during reaction of catechol hindered-phenyl phosphite (1.25×10^{-2}M) with tetralin hydroperoxide (0.25M) in tetralin at 50°C. △ in nitrogen; ○ in air.

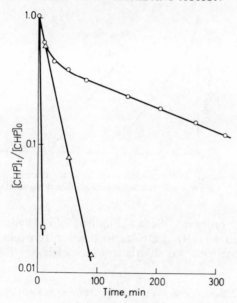

Figure 8. First order plot for the decomposition of 0.2M cumene hydroperoxide in cumene (in nitrogen) at 75°C by 0.02M catechol phosphites. ○ hindered-phenyl phosphite; △ isopropyl phosphite; □ phenyl phosphite.

from the phosphite which must be the effective catalyst for hydroperoxide decomposition. This was confirmed by adding more hydroperoxide at the end of the experiment when it was found (see *Figure 10*) that the rate of hydroperoxide decomposition was the same (within experimental error) as

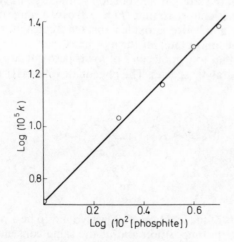

Figure 9. Logarithmic plot of pseudo first order rate constants for the catalytic decomposition of cumene hydroperoxide by catechol hindered-phenyl phosphite in chlorobenzene at 75°C.

169

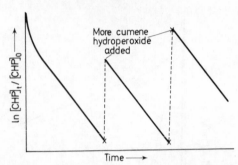

Figure 10. Catalytic activity of the products formed in the reaction between cumene hydroperoxide and catechol hindered-phenyl phosphite.

the original first order rate. Several hundred fold of hydroperoxide could be destroyed in this way. By contrast, the reaction of triphenylphosphite with cumene hydroperoxide was strictly stoichiometric (see *Table 2*).

Table 2. Stoichiometry of the reaction between triphenylphosphite and cumene hydroperoxide

$[(PhO)_3P]$ (mol. l^{-1})	[CHP] destroyed (mol. l^{-1})	$(PhO)_3P/[CHP]$
0.0235	0.025	1/1.06
0.0478	0.053	1/1.07
0.1003	0.104	1/1.03
0.2050	0.200	1/1.02

Extrapolation of the straight line portion of the first order plot of hydroperoxide decomposition to zero time (*Figure 8*) over a range of concentration ratios of phosphite and hydroperoxide indicated a constant stoichiometry of 4:1 (hydroperoxide:phosphite) for the non-linear part of the decomposition curve, suggesting that four molecules of hydroperoxide are involved in the formation of the catalytic species. The phosphate ester (III) was identified as

III

one of the products of the reaction. This was found to be a powerful catalyst for hydroperoxide decomposition and at the same concentration gave the same pseudo first order rate constant as the phosphite (*Figure 11*). Triphenyl phosphate on the other hand was inert as anticipated. In agreement with

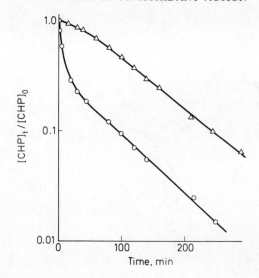

Figure 11. First order plot for the decomposition of 0.2 M cumene hydroperoxide in chlorobenzene at 75°C by: ○ 0.04 M catechol hindered-phenyl phosphite; △ 0.04 M catechol hindered-phenyl phosphite.

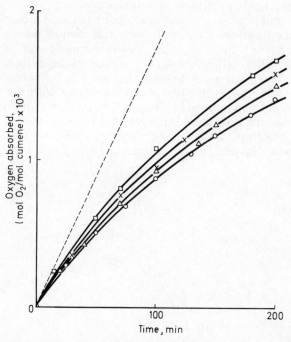

Figure 12. Inhibition of the cumene hydroperoxide-initiated oxidation of cumene in oxygen at 75°C, by catechol hindered-phenyl phosphate. [CHP]$_0$ 0.2M; ○ 0.005M; △ 0.01M; × 0.02M; □ 0.03M. (All curves are averages of four runs each.) – – – – no phosphate.

171

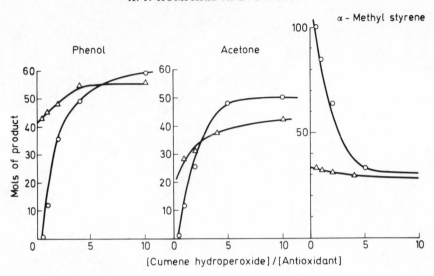

Figure 13. Cumene hydroperoxide decomposition products in chlorobenzene at 75°C. (Constant 0.2M hydroperoxide.) ○ Phosphite; △ phosphate.

this behaviour, the catechol hindered-phenyl phosphate is an effective anti-oxidant (*Figure 12*) but unlike the phosphite does not give an initial pro-oxidant effect and it is much slower in achieving full inhibition, particularly at higher concentrations due to the induction period before the active catalyst is produced. For this latter reason it appears unlikely that the phosphate is involved in the sequence of reactions leading to the antioxidant.

The antioxidant is a powerful Lewis acid as is shown by the decomposition products formed from cumene hydroperoxide by both the catechol hindered-phenyl phosphite and phosphate (*Figure 13*). At low hydroperoxide to phosphite ratios the product is primarily that expected from a homolytic breakdown of hydroperoxide, namely α-methyl styrene. This accords with the initial pro-oxidant effect on this system (see *Figure 3*) due to the formation of free radicals.

172

The catalyst which is formed acts predominantly, although not exclusively, as a Lewis acid (*Figure 13*). This dual characteristic is shown by the effect of the cumene hydroperoxide phosphite system in the polymerization of styrene. *Figure 14* shows that whereas cumene hydroperoxide alone exhibits the well known dependence of the rate of polymerization on the square root of the initiator concentration (lowest curve), there is no such simple

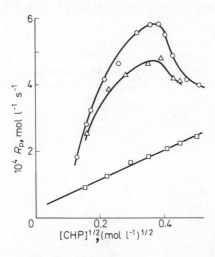

Figure 14. Dependence of rate of polymerization of styrene in oxygen on hydroperoxide concentration in the presence and absence (□) of catechol phosphites. ○ 0.01 M catechol hindered-phenyl phosphite; △ 0.01M catechol isopropyl phosphite.

relationship in the presence of the catechol phosphites [II(a) and (c)]. Both systems show a maximum rate of polymerization and hence of radical formation at the same hydroperoxide:phosphite ratio. This is consistent with the view already expressed that the initial reaction between phosphite and hydroperoxide involves the formation of free radicals. This reaction has been found to be first order with respect to each component and radical generation in this stage is consistent with the mechanism proposed by Pobedimskii[16] who postulated a bimolecular reaction between hydroperoxide and phosphite to give a radical cage structure from which the radicals may escape to initiate free radical chain reactions or may undergo further reaction within the cage to give the ultimate reaction products. Measurement of the rate of the styrene initiation process indicates that escape from the cage is a minor reaction since the efficiency of radical generation is between five and ten per cent. The main rection leads to the formation of other products by a non-radical process and it is one or more of these products which lead to the formation of a powerful Lewis acid species which inhibits both autoxidation and styrene polymerization in the later stages of the reaction.

$$(RO)_3P + R'OOH \rightarrow [(RO)_3P ---O—R']$$
$$| \quad | $$
$$OH$$

$$(RO)_3P{=}O \qquad \leftarrow [(RO)_3\dot{P}—OH\ \dot{O}R]$$
$$+ ROH$$
$$\Big| (<10\%)$$
and further
oxidation products $\qquad (RO)_3\dot{P}\ OH + \dot{O}R$

As has already been indicated, the phosphate (III) is almost certainly not the powerful antioxidant species, although it is quite as effective as an antioxidant for polypropylene as the parent phosphite ester and both are considerably more effective on a weight basis than one of the best commercial

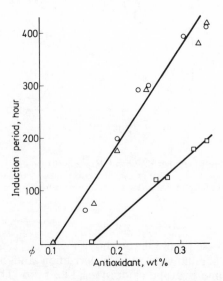

Figure 15. Induction periods to oxidation of polypropylene at 150°C stabilized by: ○ catechol hindered-phenyl phosphite; △ catechol hindered-phenyl phosphate; □ 3:1 of DLTP: Irganox '1010'.

synergistic stabilizing systems (see *Figure 15*). Other products identified in the reaction, the catechol hydroxy phosphate (IV) and its hydrolysis product (V), appear to be much more likely candidates.

IV

V

174

The former destroys hydroperoxides at a faster rate than any of the other products identified from this reaction (see *Table 3*) and, more important from the point of view of the mechanistic reaction sequence, unlike the catechol hindered-phenyl phosphate (III) neither IV nor V exhibits an initial slow stage before the onset of the pseudo first order reaction.

Table 3. Pseudo first order rate constants for the decomposition of cumene hydroperoxide by catechol hindered-phenyl phosphite II(c) and derived products in chlorobenzene at 75°C
$[CHP]_0 = 0.2M$, $[P]_0 = 0.02M$

p k_1, S^{-1}	II(c) 1.1×10^{-4}	III 1.2×10^{-4}	IV 84×10^{-4}	V 2.5×10^{-4}

In confirmation of this both the hydroxyphosphate (IV) and its hydrolysis product (V) which are formed as minor byproducts from the oxidation of the catechol phosphites have been found to be powerful antioxidants in model systems. *Figure 16* shows that the catechol hydroxy phosphate (IV) causes very rapid cessation of oxygen absorption in cumene initiated by cumene hydroperoxide and not only is no pro-oxidant effect evident but an immediate gas evolution occurs at a more rapid rate than in the case of the catechol hindered-phenyl phosphite at the same molar concentration (see

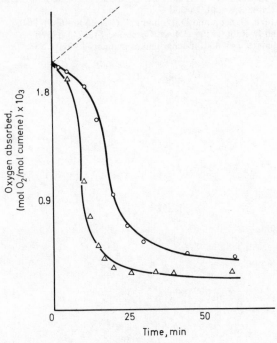

Figure 16. Gas evolution during the reaction of 0.02M catechol hydroxy phosphate with 0.2M cumene hydroperoxide in cumene at 75°C. ○ in oxygen; △ in nitrogen. ---- no phosphate.

175

Figure 3). This additive is also an effective thermal and u.v. stabilizer for polypropylene as might be expected from its peroxide decomposing activity. Comparative studies of the effects of these compounds as u.v. stabilizers will be discussed in another paper at this conference.

The authors are indebted to Monsanto (UK) Ltd for a grant to carry out this work.

REFERENCES

1 G. Scott, *Atmospheric Oxidation and Antioxidants*, Elsevier: Amsterdam (1965).
2 S. L. Fitton, R. N. Haward and G. R. Williamson, *Brit. Polym. J.* **2**, 217 (1970).
3 J. Voigt, *Die Stabilisierung der Kunststoffe gegen Licht und Narme*, p 316. Springer-Verlag: Berlin (1966).
4 E. N. Matveeva, P. A. Kirpichnikov, M. Z. Kremen, N. A. Obolyaninova, N. P. Lazareva and L. M. Popova, *Soviet Plastics*, **2**, 39 (1965).
5 P. I. Levin, P. A. Kirpichnikov, A. F. Lukovnikov and M. S. Khloplyankina, *Polym. Sci. USSR*, **5**, 214 (1964).
6 P. A. Kirpichnikov, L. M. Popova and P. I. Levin, *Trans. Kazansk. Khim.-Tekhol. Inst.* **33**, 269 (1964).
7 P. A. Kirpichnikov, N. A. Mukmeneva, A. N. Pudovik and L. M. Yartseva, *J. Gen. Chem. USSR*, **34**, 1693 (1964).
8 P. A. Kirpichnikov, L. V. Verizhnikov and L. C. Angert, *Trans. Kazansk. Khim.-Tekhnol. Inst.* **33**, 287 (1964).
9 P. I. Levin, *Soviet Rubb. Technol.* **26**, 19 (1967).
10 D. L. Rysavy and S. Slama, *Chem. Prumysl*, **18**(1), 20 (1968).
11 D. L. Rysavy and S. Slama, *Angew. Makromol. Chem.* **9**, 129 (1969).
12 S. I. Bass and S. S. Medvedev, *Russ. J. Phys. Chem.* **36**, 1381 (1962).
13 M. S. Khloplyankina, O. N. Karpukhir, A. L. Buchachenko and P. I. Leerm, *Nefta Khimiya*, **5**, 49 (1965); *Chem. Abstr.* **62**, 14435d.
14 J. D. Holdsworth, G. Scott and D. Williams, *J. Chem. Soc.* 4692 (1964).
15 C. Walling and R. Rabinowitz, *J. Amer. Chem. Soc.* **81**, 1243 (1959).
16 D. G. Pobedimskii and A. L. Buchachenko, *Bull. Acad. Sci., USSR, Div. Chem. Sci.* 1125 (1968).

THE FORMATION AND SIGNIFICANCE OF POLYCONJUGATED SYSTEMS IN THERMAL AND THERMO-OXIDATIVE PROCESSES

A. A. BERLIN

Institute of Physical Chemistry, Academy of Sciences, Moscow, USSR

ABSTRACT

The processes of formation of polyconjugated fragments during thermal oxidation of polymers and the reactions of low molecular compounds yielding polymers with an extended system of conjugation are considered. It is shown from the experimental data that the polyconjugated system (PCS) formed during thermal oxidation and thermolysis exerts initiating or activating action during the process. The cause of such properties of the PCS is the existence in macromolecules of stable biradicals capable of complex formation with diamagnetic molecules. In this complex the radical reactivity is enhanced due to an increased probability of S–T transfer (effect of local activation). Initiation of hydroperoxide decomposition caused by the PCS explains the high synergetic action of mixtures of polyconjugated polymers formed during thermal oxidation or thermolysis of monomeric inhibitors (bisphenols, aromatic amines, etc.) and initial monomers not subjected to such treatment. It is shown that during thermolysis along with paramagnetic centres (PMC), the active radicals are formed due to the rupture of valence bonds. These σ-radicals stabilized by a rigid matrix diminish the thermal-oxidative stability when molecular mobility is 'frozen'. The activity of σ-radicals, under proper conditions, can suppress the inhibiting effect of stable PMC complexes and cause transition from oxidation to burning. The probability of purposeful regulation of PCS action is associated with the information about the PMC formation mechanism during the growth of polyconjugated chain.

The author's attitude concerning the existence of paramagnetic branched chains resulting from local π-bonds unpairing in PCS is considered. Due to this fact a paramagnetic avalanche is formed because of S–T transition activation in a complex with other macromolecules by PMC formed during the preceding step. The advanced theory is in excellent agreement with known experimental data and may be useful for interpretation and prediction of the results of future investigations.

By means of thermodynamic calculations it is easy to show that in many cases the propagation of a polyconjugated chain is the most probable process in the thermal and thermo-oxidative destruction of organic compounds[1]. The reason for this is connected with the decrease in internal energy for the transition from 'saturated' compounds to substances with a system of

177

conjugation where π-electrons have the property of intra intermolecular delocalization[2]. For an appropriate length of effective conjugation, free spins appear in the latter compounds which have a great influence on the physicochemical properties of polyconjugated systems ('the effect of local activation'—ELA)[3].

In spite of 'thermodynamic advantages' for the transition to polyconjugated systems (PCS) and eventually to graphite, the rate of this process may be low on account of the real kinetic conditions. For example, although the chemical transformation

$$\sim CF_2 - CF_2 \sim \ \rightarrow C_{graf} + CF_4$$

occurs with a large negative ΔF (38 kcal/mol) under real conditions, thermolysis and radiolysis of polytetrafluoroethylene lead mainly to the formation of monomer.

Nevertheless, real kinetics cannot exclude absolutely the process which is advantageous from the thermodynamic point of view. Therefore, even in these cases when for given conditions other processes of destruction take place and PCS form with low rate, we cannot neglect their influence. It is especially true because PCS have a high inhibiting and catalytic activity in radical and some other reactions as has been proved in our research[3-6] and lately in the works of some other authors[6, 7].

In conformity with the problem of stabilization of low molecular and polymer compounds with respect to thermal, photo and radiation effects, the knowledge of the mechanism of PCS formation, their properties and the influence on the substrate transformation under the given conditions, is undoubtedly the main clue to the solution of the scientific and applied problems. It is clear that the development of these ideas is also very important for the directed synthesis of highly thermostable and other polymers.

Thus, it is easy to postulate the importance of the chemistry of polymers with the system of conjugation as the basis of both the synthesis of polymers with specific electrophysical properties and the field of obtaining polymer substances which can be used under the increasingly complicated conditions of modern techniques.

Although the ideas formulated above have been repeatedly expressed and experimentally proved since 1950, they are not taken into account in those works which neglect progress in the chemistry of polyconjugated systems.

Therefore we shall briefly consider some data illustrating the specificity of the physicochemical properties of PCS.

PARAMAGNETIC PROPERTIES OF POLYCONJUGATED POLYMERS AND THEIR INFLUENCE ON THE PHYSICOCHEMICAL PROPERTIES OF SUBSTANCES FORMING COMPLEXES WITH THEM

At the present time it is well known that polymers with the π-π and p-π-systems of conjugation have rather stable paramagnetic centres (PMC) in condensed phase and in solutions with a narrow EPR singlet and a free-electron g-factor. It is essential that PMC should exert influence on the physical and physicochemical properties of PCS including dark and photo-induced conductivity, absorption spectra, luminescence and reactivity. This influence on the properties of conjugated substances forming complexes with PMC was discovered in 1962 in our laboratory and called 'the effect of local activation'.

At the present time two assumptions about the nature of paramagnetism in PCS are of a great importance among others:

(1) the biradical hypothesis[3, 8–10] which considers PMC as a result of the local decay of π-bonds in polyconjugated macromolecules with the longest chain. With respect to this hypothesis S–T transition may be realized due to thermal excitation and the resulting triplet can be transformed into a biradical. It has also been assumed that the probability that these transitions will occur is connected with a small value of the energy gap in packing like associates of fractions of the polymer homologues with long conjugated chains, with the breach of coplanarity and the gain in resonance energy due to delocalization of the unpaired spin along the system of conjugation.

(2) the hypothesis about the predominant role of electron charge transfer based on the assumption of the presence of close-situated energy levels in PCS. According to this hypothesis charge transfer proceeds via one-electron intermolecular transfer leading to the formation of a stable ion–radical complex[11, 12].

The second hypothesis cannot be considered as the foundation of the general theory of PCS paramagnetism because it does not explain some experimental facts[13] including the absence of the contribution of dielectric relaxation in polymer hydrocarbons with a system of conjugation[10].

Moreover, according to quantum-chemical calculations, the one-electron transfer is of low probability in non-polar PCS even for chains of infinite length.

The last remark is also true, although to a much smaller degree, for PCS triplet excitation which is the base of the biradical hypothesis.

We suppose that the general theory for paramagnetism of PCS may be worked out by means of a model of the paramagnetic chain processes ('paramagnetic chains') which will be considered below. According to this model the triplet excitation is a primary act leading to the formation of stable biradicals or ion–radicals (polar media). These processes proceed with low probability at reasonable values of the PCS energy gaps. For example, the probability of occurrence of a triplet population does not exceed 10^{-15} and 10^{-22} at room temperature and at the energy of S–T transition, $E_{ST} = 20$ to 40 kcal/mol. The resulting PMC are rather stable macromolecular compounds with free spins delocalized along extensive blocks of conjugation.

In real systems macromolecules of PCS are bound in π-complexes and so the resulting free spins interact also with π-electrons of the complex-forming chains, but with a smaller exchange frequency. In these 'complexes with spin transfer'[2], the probability of S–T transition increases and the value of the energy gap apparently decreases.

At corresponding energy expenditures this fact leads to the decay of the π-bond in such complexes under the influence of the primary spin, i.e. to the effect of local activation.

As a result one spin initiates the formation of two new ones. This leads

179

to the propagation of a greatly branched chain process. We called this process 'branched paramagnetic chain'.

Owing to the branched-chain character of the reaction and to the high value of the pre-exponential factor ($\geqslant 10^{13}$ to 10^{15} 1/s), the process rate steadily increases until it becomes a 'paramagnetic' avalanche; as a result stable PMC are accumulated, which allows their detection by EPR at rather low temperatures.

It is necessary to note the existence of a relationship between the rate of formation, the nature of the polymer and the kinetics of paramagnetic chains.

As PMC are accumulated, two processes take place: the first is PMC 'recombination' with the formation of spin–spin complexes[14] and the second is PMC decay, i.e. chain termination because of polyconjugation cleavage and spin inactivation in reactions with admixtures.

Disregarding the kinetics of PCS formation the general data about branched paramagnetic chains may be represented by the following simplified scheme:

The formation of primary PMC

$$\downarrow P\uparrow \overset{k_i}{\underset{E}{\rightarrow}} \uparrow P\uparrow \tag{1}$$

The formation of a complex with spin transfer and the propagation of the branched chain

$$P\uparrow + \uparrow P_1\downarrow \overset{k_s}{\underset{}{\rightleftharpoons}} [P\{\ \}P_1\uparrow] \overset{k_a}{\underset{E}{\rightarrow}} P\uparrow + \uparrow P_1\uparrow \tag{2}$$

$$P_1\uparrow + \uparrow P_2\downarrow \overset{k_s}{\underset{}{\rightleftharpoons}} [P\{\ \}P_2\uparrow] \overset{k_a}{\underset{E}{\rightarrow}} P_1\uparrow + \uparrow P_2\uparrow \quad \text{etc.} \tag{3}$$

The deceleration of the propagation due to spin–spin complex formation

$$\uparrow P_1\uparrow + \uparrow P_2\uparrow \overset{k_b}{\underset{}{\rightleftharpoons}} [\uparrow P_1\{\ \}P_2\downarrow] \tag{4}$$

The termination of the paramagnetic chain due to the cleavage of conjugation or spin inactivation

$$P_i\uparrow + x \overset{k_t}{\rightarrow} P_ix \tag{5}$$

In this scheme the following symbols have been used: P_i are molecules of PCS with i links in the block of effective conjugation, arrows $\uparrow\downarrow$ and $\uparrow\uparrow$ mark the ground and triplet states (or stable biradical) correspondingly; x denotes molecules of the admixture which are capable of reacting with the spins of a stable radical; k_i, k_s, k_a, k_b, k_t are rate constants of the formation of primary centres, the formation of the complex 'spin–π-bond', the formation of biradicals under local activation (reaction 2), deceleration and decay of PMC, respectively.

In the first approach, the kinetics of the PMC accumulation may be expressed by the equation

$$d[\text{PMC}]/dt = k_i[\text{PCS}] + k_a[\text{PMC}][\text{PCS}] - k_h[\text{PMC}]^2 - k_t[\text{PMC}]x \tag{6}$$

where $k_h = K \times k_a$, and K is the equilibrium constant of the spin complex (reaction 4). Under the quasi-stationary conditions ($d[\text{PCS}]/dt = 0$), integration of equation 6 gives:

$$[PMC] = \frac{(\gamma/a)^{\frac{1}{2}}\{c + \exp[-2(\gamma a)^{\frac{1}{2}}t]\}}{c - \exp[-2(\gamma a)^{\frac{1}{2}}t]} + \frac{b - d}{2a} \tag{7}$$

where: $a = k_b$; $b = k_a[PCS]$; $\gamma = k_i[PCS] + (b - d)^2/4a$; $d = k_t[x]$; and c is the constant of integration which can be determined from the initial conditions.

The analysis of expressions 6 and 7 shows the following features:

(1) The curve of PMC accumulation has an **S**-type form typical of auto-catalytic processes.

(2) At $t \to \infty$, the PMC concentration reaches the limiting value $[PMC]_\infty$ $= (\gamma/a)^{\frac{1}{2}} + (b - d)/2a$ which does not depend on the initial concentration of PMC, i.e. $[PMC]_0$.

(3) The spin accumulation rate increases in the region of autoacceleration with increasing $[PMC]_0$, i.e. the effect of local activation becomes apparent in the process.

(4) The linear (or bimolecular) termination (reaction 5) leads to an increase in the induction period and to a decrease in the limiting value of the PMC concentration.

(5) The decrease of the deceleration rate (reaction 4) must lead to an increase of $[PMC]$ and to a decrease in the number average molecular weight with increase in temperature. Indeed, under such conditions the equilibrium constant of the spin complex decreases, and consequently, cracking of such quasi-molecules with the formation of stable radicals takes place.

(6) If the reaction of chain termination (reaction 5) contributes markedly to the process, then other things being equal, the rate of PMC accumulation and its temperature dependence depend on the relation between the deceleration and termination rate constants.

The ideas presented above even in an approximate form are in good agreement with well-known experimental facts. For example, it has been shown that the kinetics of PMC accumulation for PCS synthesis of polymer thermolysis is really described by **S**-shaped autocatalytic curves which go to a limit. This fact is illustrated by *Figures 1* and 2[15, 16] and is in complete

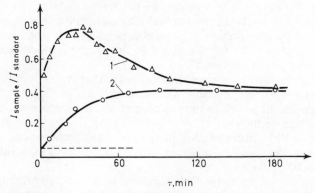

Figure 1. The dependence of the relative e.p.r. signal for thermal polymerization of phenylacetylene on time τ(min): (1) in resonator; (2) after cooling in air for five minutes.

181

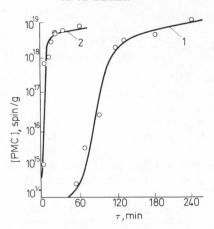

Figure 2. The kinetics of PMC accumulation in anthracene thermolysis at $T = 450°C$ (vacuum 10^{-5} mm Hg): (1) anthracene; (2) anthracene with addition of one per cent of polyanthracene containing 5×10^{17} spin/g.

agreement with our theory. *Figure 2* illustrates also the activation of anthracene thermolysis by the paramagnetic fraction. As we can see, the induction period of PMC accumulation for anthracene dehydropolymerization may be practically excluded due to ELA.

At the present time, a great number of experimental facts exist to demonstrate that ELA appears over a wide range of processes and phenomena such as thermal, thermo- and photo-oxidizing destruction of polymers[4], catalysis of radical polymerization and some other polyreactions[14, 15, 17], the activation of *cis–trans* transition[21] and some other organic reactions[22, 23]. The effect of local activation is also reflected in the physical properties of the substance. It affects the activation energy of PCS electroconductivity, the changes of the lifetime of the photocurrent carrier, the value of fluorescence yield, and some other properties[24, 25].

Recently, the influence of spin on the reactivity of stable nitrogen oxyradical[26] has been detected experimentally, along with the catalytic influence of oxygen on the destruction of aromatic polyamides[27] and oxygen and nitric oxide catalysis of CCT formation in the interaction between chloroanhydride with tertiary amines[28].

Apparently, in two latter cases the ELA is realized due to spin complexes of the substrate with such stable gas phase radicals as O_2 and NO. It is possible that the ELA will change our views about bioprocesses, and also about oxidation–reduction.

The limit of PMC accumulation predicted by our theory is confirmed by both quantum-chemical calculations and experiments. For soluble PCS it is equal to 10^{20} spin/g.

The possibility of reversible PMC accumulation and its correlation with the decrease of M_n has been proved experimentally in our work[29] and is shown in *Figure 3*. Apparently, under these conditions the decay is intensified

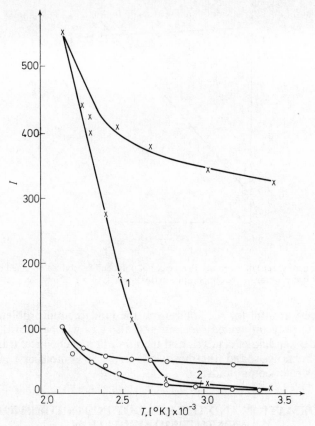

Figure 3. The dependence of e.p.r. intensity of a narrow singlet on inverse temperature for poly-phenylacetylene obtained by catalytic and thermal polymerization: (1) polyphenylacetylene obtained by polymerization in the presence of $(C_2H_5)_3AlTiCl_3$ at $T = 20°$ to $50°C$; (2) poly-phenylacetylene obtained by thermal polymerization at $T = 150°C$.

as might be expected. This process is more intensive for PCS obtained at a smaller viscosity (i.e. in solutions) and at lower temperatures.

The fact that the reversibility of PMC concentration is not complete may be explained by the structural peculiarities of PCS, causing a high value of the relaxation time[30].

At the present time we have only preliminary experimental data confirming the predictions of the theory about the reversible changes in the number average molecular weight of PCS with moderate heating. Nevertheless, the comparison of this process proceeding in air and in a vacuum (see *Figure 4*) clearly shows the role of PMC inactivation by oxygen preventing spin–spin complex formation (see above, reaction 4). This makes it possible to observe the decrease in M_n determined at room temperature.

It seems to us that all that has been described above shows that the combination of the effect of local activation with Semenov's theory of branched

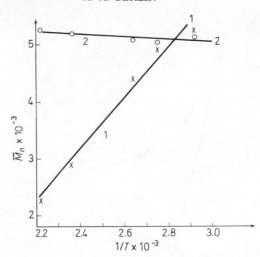

Figure 4. The dependence of the molecular weight of polyphenylacetylene obtained by catalytic polymerization on the temperature of treatment: (1) in air, (2) in a vacuum.

chains is very fruitful for the solution of the fundamental problems in the chemistry of polyconjugated systems, and also—as will be shown below— for the understanding of catalysis and the inhibition of chemical transformations in thermolysis and thermo-oxidation of low molecular and high molecular weight compounds.

THE FORMATION AND THE ROLE OF PCS IN THERMOLYSIS AND THERMO-OXIDATION

As has been shown above, some finite probability exists of transforming organic substances into polyconjugated polymer compounds in the case of an appropriate thermal action. This transformation has been investigated for benzene and some aromatic hydrocarbons with condensed rings (anthracene, bianthryl, pyrene, etc.). For example, the autocatalytic accumulation reaction of paramagnetic oligomers ($M_n = 1200$–14000). $C_{PMC} = (3$–$8) \times 10^{17}$ spin/g with anthronylene links proceeds in anthracene thermolysis in a vacuum or in an inert atmosphere at $T = 400°$ to $450°C$. The investigation of the chemical mechanisms favours Scheme I in this process:

Scheme I

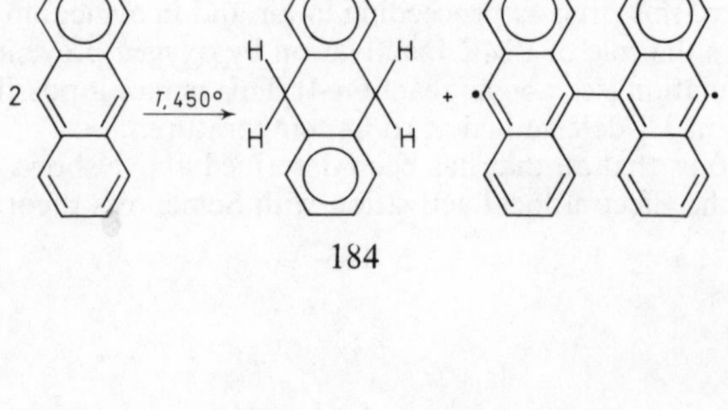

184

PMC accumulation in the polymer activates the process. This has been proved by the vanishing of the induction period at increasing rates of accumulation due to the introduction of catalytic amounts (up to one per cent) of the soluble polyanthracene paramagnetic fraction.

The investigations[15] carried out show that the increase in the polymer yield and PMC concentration in the polymer lead to an increase in the gaseous reaction products (H_2, CH_2, C_2H_6, C_3H_8, C_2H_4, C_4H_8). Apparently, complex reactions of anthracene cracking accompanied by dehydro-anthracene dehydration are activated as a result of the effect of local activation.

Electron-donor properties and high radical reactivity of paramagnetic PCS make them able to react with oxygen and radicals, and also to initiate hydroperoxide decay without any yield of radical products[2, 23]. Therefore, even small amounts of PCS formed during thermo-oxidation lead to the braking of this process (the effect of 'self-stabilization'). Thus, for example, it is known that thermo-coloured resin-like products formed by the oxidation of lubricating oils are strong inhibitors of their oxidation[31].

The inhibiting activity of resin-like oxidation products has been described in ref. 32 for dibenzyl thermo-oxidation. The kinetics of 'self-braking' for slow chain reactions of resin-like oxidation products forming in this process has been considered in detail by Emanuel and co-workers[33, 34].

The authors show that the forming resins inhibit thermo-oxidation of butane and methyl ethyl ketones (see *Figure 5*). By reason of the correlation between the rates of the decrease in acetaldehyde yield and the increase in resin formation, the authors assume that the inhibiting substance is formed by acetic anhydride polycondensation leading to PCS formation. In connection with this fact, the results of ref. 34 are of great interest. As has been shown in this work, self-braking of propylene oxidation in the presence of acetic anhydride is connected with resin formation. These resins are polyconjugated

185

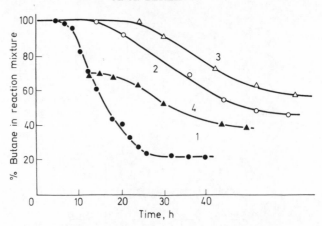

Figure 5. The influence of resin additions on the kinetics of butane expenditure in thermo-oxidation (145°C, 50 atm): (1) without additions; (2) 0.9 g of resin per 1 mol of butane added at the beginning of the process; (3) the same for 1.8 g of resin; (4) 0.9 g of resin per 1 mol of butane added 12 h after the beginning of the process.

oligomers with a narrow e.p.r. singlet typical of PCS and with PMC concentration 10^{17} spin/g.

Unfortunately, the chemical mechanism of resin formation, typical of the majority of thermo-oxidation processes ($T = 100°$) of low molecular substances, has not so far been investigated in detail. However, we may agree with the authors referred to above that carbonylic compounds forming in the hydroperoxide decay, which proceeds in the presence of metals with variable valence particularly readily, take part in these processes. Scheme II may be proposed as a hypothesis.

At the present time one may consider the existence of two types of paramagnetic particles in the thermolysis of organic polymers at $T > 300°C$ to be proved. These types are: (a) stable PMC of biradical nature and forming complexes with diamagnetic chains; (b) rather active σ-radicals, appearing as a result of bond cleavage in the main chain.

Other things being equal, the ensuing fate of these particles depends on the chemical nature, structure and submolecular organization of the polymer. For very rigid systems σ-radicals may be accumulated in a large amount because the chemical relaxation of the system leading to their decay (i.e. the combination of σ-radicals) plays a small part. Such a situation is advantageous for the thermal stability of the system, because the decay of lightly diffusing carriers of radical destruction will be a result of their interaction not only with PMC but also with the 'stuck' σ-radicals. However, another situation arises for systems where oxygen is present. In this case, for a small chemical relaxation rate, free σ-radicals interact strongly with oxygen with a large thermal effect (without activation energy) and may not only accelerate thermo-oxidation, but also stimulate thermal ignition of the system at corresponding concentrations.

The experimental data obtained in our laboratory are in complete agreement with all aforesaid and indicate that a trivial cage-effect cannot be

186

Scheme II

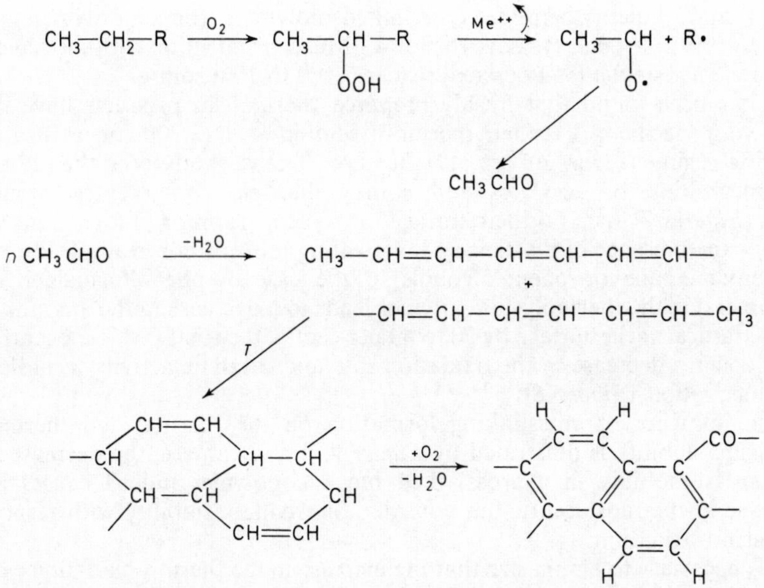

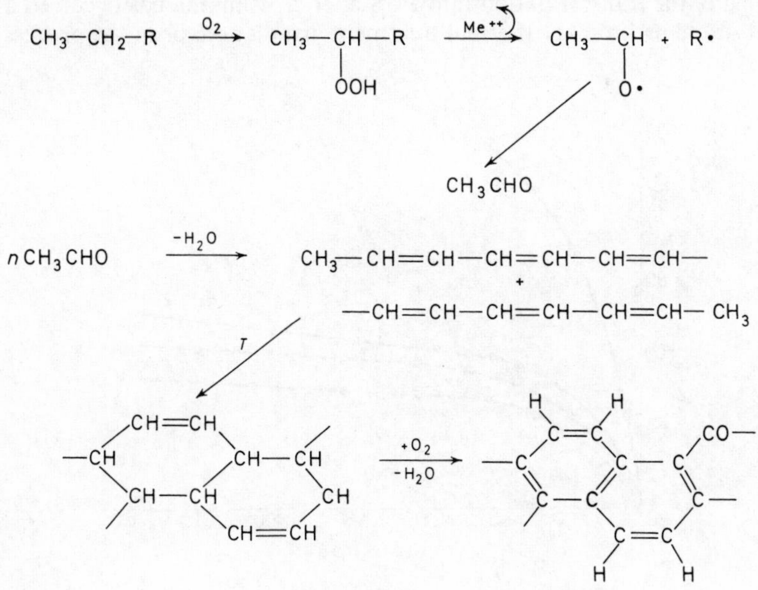

$$CH_3-CH_2-R \xrightarrow{O_2} CH_3-\underset{\underset{OOH}{|}}{CH}-R \xrightarrow{Me^{++}} CH_3-\underset{\underset{O\cdot}{|}}{CH} + R\cdot$$

$$CH_3CHO$$

$$n\,CH_3CHO \xrightarrow{-H_2O} CH_3-CH=CH-CH=CH-CH=CH-$$
$$+$$
$$-CH=CH-CH=CH-CH=CH-CH_3$$

$$CH_3-CH_2-R \xrightarrow{O_2} CH_3-\underset{\underset{OOH}{|}}{CH}-R \xrightarrow{Me^{++}} CH_3-\underset{\underset{O\cdot}{|}}{CH} + R\cdot$$

$$CH_3CHO$$

$$n\,CH_3CHO \xrightarrow{-H_2O} CH_3-CH=CH-CH=CH-CH=CH-$$
$$+$$
$$-CH=CH-CH=CH-CH=CH-CH_3$$

187

applied to PCS or to polymers transforming into these substances[35, 36]. For example, it was found that processes leading to the formation of thermo-coloured polymers with fragments of condensed aromatic compounds take place in vacuum thermolysis (10^{-3} to 10^{-5} mm) of phenolformaldehyde resins and *p*-diethylbenzene crosslinked polymers (or co-polymers) at 450° to 500°C. In such systems, the total PMC concentration with the typical narrow e.p.r. singlet (3–4 oersted) reaches 10^{20} to 10^{21} spin/g.

It has been found that freshly prepared thermolysis products have the following specificity: (1) their thermo-oxidation at $T = 300°$ turns into an ignition régime (*Figure 6*)[36, 37]; (2) they are effective catalysts of the radical polymerization of acrylates and some other electron-acceptor mono-mers (*Figures 7, 8*)[20]; (3) their storage in oxygen, argon or in a vacuum, as well as the presence of plasticizers inactivating radicals (for example, dioxy-diphenylmethane or phenolalcohols in the case of phenolformaldehyde oligomers) in the thermolysing polymer leads to a decrease in the amount of active radical at the surface by 30 to 40 per cent of the total PMC concentra-tion, a sharp decrease in the oxidation rate and catalytic activity in radical polymerization (*Figure 8*)[20, 36–48].

The influence of crosslinking formation on the thermal and thermo-oxidizing stability is illustrated by *Figure 9*. As we can see, the decrease in molecular mobility in a crosslinked block copolymer indeed causes an increase in thermostability, but considerably reduces stability with respect to thermo-oxidation.

It is necessary to emphasize that the increase in the thermo-oxidation rate and even the transition to thermal ignition at $T \geqslant 300°$ to 400°C and an increase in oxygen pressure are also possible for fusible and soluble polymers capable of the transformation into PCS after crosslinking has occurred as a result of the preliminary thermal treatment, irradiation, or in the process of

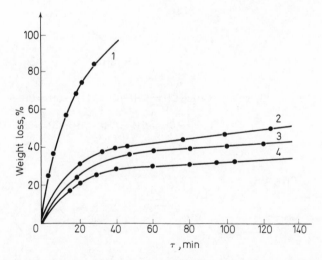

Figure 6. Thermo-oxidizing destruction of thermolysed resin at 300°C ($p = 1$ atm): (1) 'fresh' thermolysed resin without phenol alcohol; (2) the same resin after storage in oxygen for 12 hours; (3) the same resin after storage in oxygen for 36 hours; (4) the same resins with phenol alcohols.

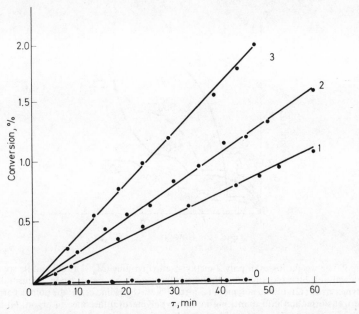

Figure 7. The dependence of the kinetics of polymerization of methylmethacrylate (MMA) catalysed by thermolysed phenolformaldehyde resin without phenol on the concentration of paramagnetic centres: (0) MMA without catalyst; (1) MMA with two per cent thermolysed resin containing 2×10^{18} spin/g; (2) the same with 1.7×10^{21} spin/g; (3) the same with 5×10^{21} spin/g.

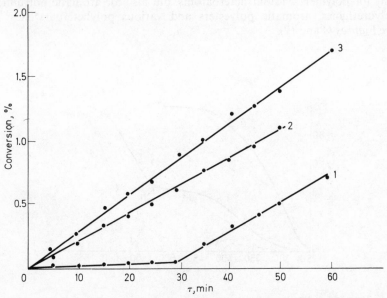

Figure 8. The change of the catalytic activity of the thermolysed resin (TR) in the polymeriza-tion of methylmethacrylate because of storage in air. The duration of keeping: (1) 60 min, (2) 30 min, (3) 0 min.

189

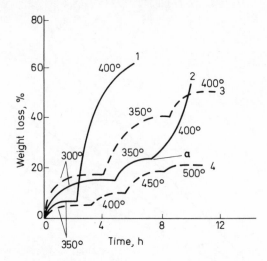

Figure 9. The weight losses of poly-2-methyl-5-ethylpyridine ($\overline{M}_n = 1\,700$) and its space cross-linked insoluble block copolymer with *n*-diethylbenzene in air and in argon: (1) insoluble block copolymer in air; (2) soluble polymer of 2-methyl-5-ethylpyridine in air; the point **a** corresponds to the transformation into unmelting insoluble polymer; (3) the same in argon; (4) insoluble block copolymer in argon.

thermo-oxidation (see *Figure 10*). All these features may be observed not only for polymers without heteroatoms, but also for aromatic polyamides, polyurethanes, aromatic polyesters and various polyheteroarylenes[39-41] (see *Figures 11* and *12*).

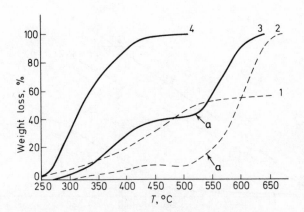

Figure 10. The comparison of thermal and thermo-oxidizing destruction of soluble polyphenylacetylene obtained at 150°C (4, 3) and 400°C (1, 2). Curves 2, 3—in air; 1, 4—in a vacuum. In the point **a** the polymer loses fusibility and solubility. Rate of temperature increase: 3C°/min.

190

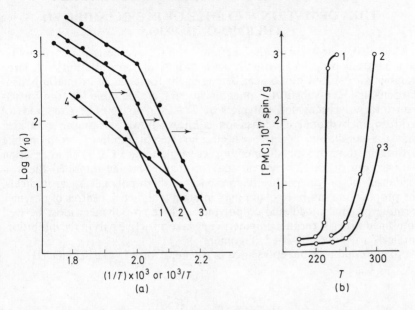

Figure 11. The kinetics of (a) polyurethane decomposition and (b) PMC accumulation: (1) polyurethane from 2,4-toluene diisocyanate(TDI) and diethylene glycol; (2) the same from TDI and propylene glycol; (3) the same from TDI and ethylene glycol (EG); (4) the same from TDI + EG with the addition of thermolysed polyurethane ($\sim 10^{17}$ spin/g). [V_{10} is the rate of decomposition of polyurethane.]

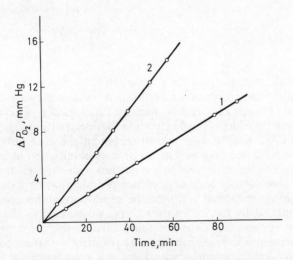

Figure 12. The influence of size of soluble fraction in polyurethane photo-oxidation based on 2,4-toluene diisocyanate and glycol containing PMC on thermo-oxidation (220°C) of polyurethane from hexamethylenediisocyanate: (1) initial polyurethane from hexamethylenediisocyanate and glycol; (2) the same with one per cent addition of soluble product of toluenediisocyanate–ethyleneglycol polyurethane (thePMC concentration is 10^{18} spin/g).

191

THE FORMATION AND ROLE OF PCS IN INHIBITED THERMO-OXIDATION

One of the most auspicious methods of PCS synthesis is dehydropoly-condensation proceeding usually with radical or ion initiation[42]. A large number of assorted monomers belonging to the class of aromatic hydro-carbons, halogen-substituted phenols, amines, etc. suitable for transforma-tion into polyconjugated polymers at $T \leqslant 200°$ to $250°C$ by methods of oxidizing dehydropolycondensation is known now. This suggests an idea that the conditions when low-molecular weight inhibitors with mobile hydrogen atoms are transformed into corresponding PCS by an interaction with oxygen or peroxy radicals may be realized in the inhibited thermo-oxidation of polymers. In the presence of high-polymer macroradicals, the propagating inhibitor chain may interact with active centres of polymer forming original inoculated co-polymers. These co-polymers must be well combined with a saturated polymer. They take an active part in the inhibition, similarly to polyconjugated oligomers[4].

All aforesaid may be represented in a general form by the Scheme III:

Scheme III

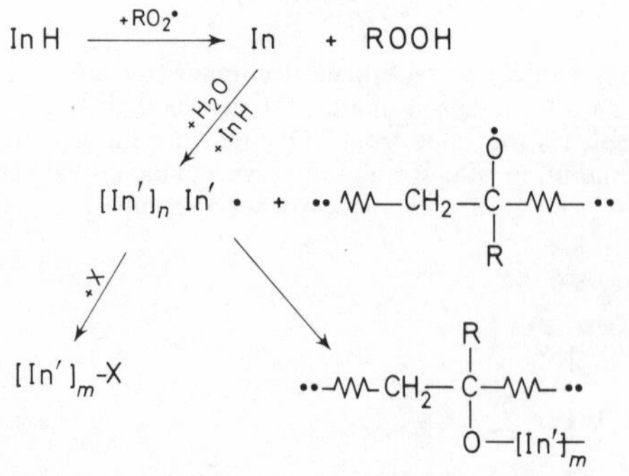

As a result of these processes, products with higher radical reactivity than initial monomer inhibitors are formed because they are characterized by a greater branch conjugation and by the presence of complex bound PMC causing the effect of local activation. Besides, polyconjugated polymers containing PMC have a lower ionization potential and so initiate the decay of hydroperoxide and take part in the process of initiation at high tempera-tures to a much lesser degree. In connection with this fact it is natural to expect an anomalously high dependence of induction periods on the con-centration of the monomer inhibitor, the appearance of the second induction period stipulated by the PCS formed, and an increase in the temperature limit for the application of the inhibitor suited for the above processes.

The experimental data are in complete agreement with these considerations and lead to some conclusions which are of a great scientific and practical interest[4].

As was emphasized in our work on the braking of the oxidation of paraffins by products of anthracene thermolysis, the oxidizing dehydropolycondensa-tion leading to the formation of coloured polymer products sharply increasing the inhibitor activity proceeds in the inhibited thermo-oxidation of ceresin ($T = 180°$ to $200°C$) in the presence of anthracene and its complexes with tetracyanoethylene. In this paper we have given evidence for the presence of polyconjugated chains and PMC in such products[43, 44].

Later a report was published[45] where the authors described high inhibiting activity of coloured compounds forming in the oxidation of p-methyl-alkylphenols and alkyl-bis-phenols in the inhibited thermo-oxidation ($T = 200°C$) of polypropylene and co-polymers of tetrahydrofuran with ethylene oxide.

Figure 13 taken from the work carried out by Levantovskaya, Kovarskaya and co-workers[46] illustrates the accumulation of coloured products of the oxidation of 4,4-methyl-bis(2,6-di-tert-butylphenol) used as the inhibitor within the induction period and the development of thermo-oxidation when the products are transformed into uncoloured substances. These results

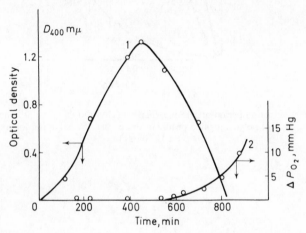

Figure 13. The kinetic curves of the changes in optical density (1) at $\lambda_{max} = 400$ mμ(D_{400}) in thermo-oxidizing destruction of tetrahydrofuran–propyleneoxide copolymer in the presence of 0.05 mol/kg 4,4'-methylene-bis(2,6-di-tert-butylphenol) antioxidant $T = 160°C$, $P_{O_2} = 200$ mm Hg; (2) corresponding change of partial pressure of oxygen.

are in complete agreement with our investigations and with the work carried out by Pospíšil and Gömöry with co-workers, where the increase in inhibiting activity and the existence of the so-called second induction period in the presence of coloured products formed from aromatic amines and alkylphenols within inhibited oxidation of ceresin, rubber, polyethylene, and some other polymers have been demonstrated (see *Figures 14, 15*)[4, 47].

The data about the synergism of the inhibiting action in mixtures of unoxidized and oxidized amines and phenol stabilizer are of great interest. Such effects have been studied by us and by the cited authors in the following systems: polyanthracene–anthracene[4, 43], polydiphenylamine–diphenylamine[4, 47, 48], the unfractionated coloured products of bisphenol and diphenyl-p-phenylenediamine oxidation separated from monomers and also for the process of thermo-oxidation of paraffins, polypropylene, rubber and polyethylene ($T = 160°$ to $200°C$).

193

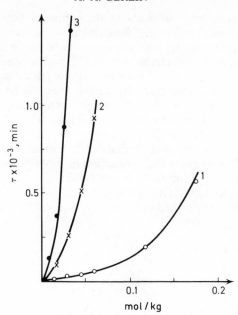

Figure 14. The induction period of ceresin oxidation ($T = 180°C$) in the presence of diphenyl-amine (DPA) (1) and polyconjugated products formed in DPA oxidation for six hours (2) and for 15 hours (3).

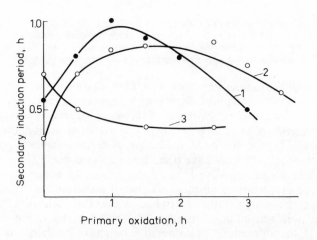

Figure 15. The second induction periods of polyethylene oxidation. Inhibitors: 1—4,4'-dihy-droxydiphenyl; 2—phenylcyclohexyl-p-phenylenediamine; 3—N,N'-diphenyl-p-phenylenedi-amine.

194

The mechanism of these processes has been investigated by some authors (for example, see refs. 46, 49, 50) using thin layer chromatography with aluminium oxide and isolation of coloured products by benzene, petroleum ester–ethylacetate with the subsequent application of e.p.r., n.m.r., i.r. and electronic spectrometry and chemical analysis.

It is interesting to note that by using these methods, the authors usually came to the conclusion that an increase in the inhibiting activity was connected with the transition to benzoquinone (phenol oxidation) or quinone-imines (aromatic amine oxidation). However, such results do not agree with Pospíšil's data which show a considerable decrease in the benzoquinone inhibiting activity in comparison with hydroquinone[51], and also with our investigations giving evidence for the formation of polyconjugated oligomers containing PMC in aromatic inhibitor oxidation[4].

We suppose that the reason for this disagreement resides in the possible isolation of low molecular products only from the absorbent, since the polyconjugated paramagnetic polymer has not been extracted. Indeed, it is well known that it is impossible to isolate the polymer product with the usual solvents for PCS absorption on aluminium or silicone oxide (see for example ref. 13). This is why it is difficult to extract oxidizing polycondensation products by using such methods of fractionation and to estimate this role in thermo-oxidation inhibition.

In connection with this fact, let us consider some data about the inhibiting action and thermo-oxidation of diphenylamine (DPA) and analogues obtained from it. The formation of PMC, the intensification of colour in the induction period and the high limit concentrations have been established by using DPA as the inhibitor of thermo-oxidation (160° to 220°C) of ceresin and polymers (rubbers, polypropylene). This justifies the assumption that DPA may be transformed into more active polyconjugated polymer products. The results of works on the isolation of polymer products of oxidation and their investigation confirmed what has been said above and showed that

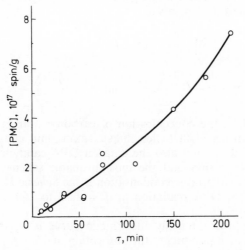

Figure 16. The increase in PMC concentration in diphenylamine oxidation (220°C, P_{O_2} = 1 atm).

195

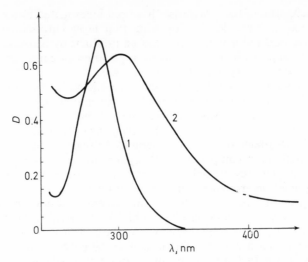

Figure 17. Absorption spectra of monomer (DPA) and polymer obtained by its oxidation (220°C, $P_{O_2} = 1$ atm): (1) DPA electronic spectrum; (2) the same for PDA with $\overline{M}_n = 2\,500$.

it was necessary to investigate the mechanism of DPA oxidation and the properties of polymer products. We investigated also some DPA analogues (phenyl-α and β-naphthylamines, alkyl and alkylene substituted DPA, diphenyl-p-phenylenediamines).

As has been found, oxidizing dehydropolycondensation accompanied by an autocatalytic increase in PMC concentration and by the formation of coloured polymer products with number average molecular weight from 540 to 2600 and stable PMC concentration from 10^{17} to $(3-8) \times 10^{18}$ spin/g takes place in DPA thermo-oxidation (see *Figures 16* and *17*). Chemical analysis, i.r. and n.m.r. investigations of the polymer formed show the presence of structural elements like

At the same time, the concentration of carbonyl quinone oxygen is very small (one atom per eight to ten links). Taking into consideration all the facts indicated above and also the data on DPA catalytic oxidation[48], we came to the conclusion that the most probable scheme of the chemical mechanism for dehydropolycondensation is the *Scheme IV*.

The kinetics of DPA oxidation at $T = 225°$ to $260°$C may be seen in *Figure 18*.

Plotted semilogarithmically the kinetic curve is a straight line till a process depth corresponding to the absorption of 0.6 to 0.7 g/mol of oxygen per 1 g/mol of DPA. This corresponds to alienation of amine hydrogen and

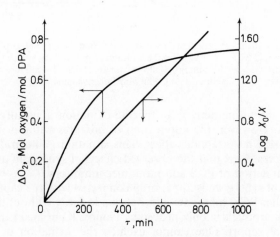

Scheme IV

Hypothetical scheme of oxidative
dehydropolycondensation of
secondary aromatic amines

Figure 18. The kinetic curves of oxygen absorption in the oxidation of diphenylamine (DPA) (225°C, $P_{O_2} = 1$ atm) (1) and its semilogarithmic anamorphosis (2). $X_0 = 0.75$ mol O_2/mol DPA; $X = (0.75 - \Delta O_2)$ mol O_2/mol DPA.

H-atom in the phenyl ring. The effective activation energy for oxidation is 20 kcal/mol.

Polyconjugated polymers isolated in this stage are easily soluble and fit in with the above-mentioned characteristics. The interaction of oxygen with H-atoms at the position ortho with respect to amine nitrogen begins to play a considerable part at deeper oxidation. As a result, the chain branches and this leads to space crosslinking. Along with this, it has been established that insoluble products are already formed in the early stages of oxidation of 4,4'-dialkyl or 4,4'-oxyderivatives of DPA proceeding at a higher rate where the oxygen attack is directed at both H-groups, hydrogen in benzene, and the alkyl groups. Analogous results have been obtained for the joint co-oxidation of DPA with its 4,4'-disubstituted derivative.

It has been found that there is a catalytic influence of PMC ([PMC] $\leqslant$ (3–5) $\times$ 10^{18} spin/g) appearing as a result of oxidizing DPA dehydropoly-condensation (polydiphenylamine—DPA) on the kinetics of its oxidation. We have demonstrated that PDA added to DPA reduces the induction period of diphenylamine oxidation (*Figure 19*). As we can see, it is necessary to take into consideration the effect of local activation in this case too.

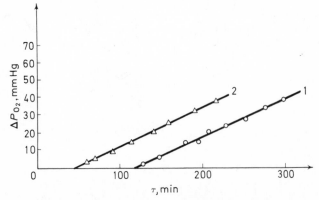

*Figure 19.*The dependence of induction period and initial rate of oxygen absorption in diphenylamine oxidation (225°C, P_{O_2} = 1 atm) on the addition of polydiphenylamine ($\overline{M}_n$ = 2000, the PMC concentration is 5 $\times$ 10^{18} spin/g). (1) diphenylamine; (2) the 1:4 mixture of polydiphenylamine ($\overline{M}_n$ = 2400) with diphenylamine.

The PDA inhibiting activity in thermo-oxidation is usually higher than that of DPA. However, the application of mixtures without monomers of PDA with DPA has a greater effect. This situation is illustrated in *Figure 20*. These curves show that the observed effects of synergism are connected with the joint action of PCS and monomer inhibitors.

The cause of synergism is apparently connected with the capacity of PDA (as well as other PCS) to initiate hydroperoxide decay without the formation of free radical products in bulk proved by Ivanov, Kobrianski and the author of the present report. One cannot exclude the formation of PMC–DPA spin complexes with the activation of their reactivity due to the effect of local activation.

It should be added that a rather small content of hydrogen of the NH-

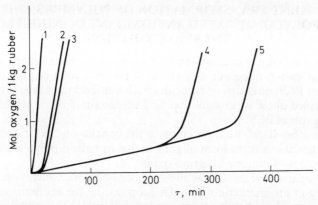

Figure 20. The kinetics of oxygen absorption with 1,4-*cis*-polybutadiene at $T = 150°C$ and $P_{O_2} = 1$ atm: (1) without inhibitor; (2) with 0.75% addition of PDA; (3) with 2.75% addition of DPA; (4) with 2.0% addition of PDA; (5) with addition of 0.75% PDA + 2.25% DPA.

group in soluble PDA or the absence of 'sticking' active radicals (see above) in such polymers practically exclude their ability to take part in the act of initiation.

Meanwhile, they have a fairly low ionization potential and high radical reactivity and so they are the effective electron donors with respect to oxygen and hydroperoxides; they destroy them and react with the products of radical decay.

The capacity of PDA to exhibit synergistic effects in mixtures with various substances including dilanoylthiodipropionate, trinonylphenylphosphite (oxidation of nitrile rubber SKN-26), anthracene (oxidation of ceresin, paraffins, polyolefins) are in complete agreement with all aforesaid (see *Figure 21*).

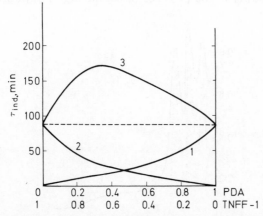

Figure 21. The dependence of induction period in nitrile rubber (SKN-26) oxidation on the content of binary mixture PDA ($\overline{M}_n = 2050$) with Polygard at constant total concentration of components. The inhibitor concentration 0.06 mol link of PDA per kg of rubber at $P_{O_2} = 760$ mm Hg, $T = 160°C$. (1) separately introduced PDA; (2) separately introduced Polygard (trinonylphenylphosphite—TNFF); (3) mixture of PDA and Polygard.

199

JOINT TRANSFORMATION OF POLYMERS AND POLYCONJUGATED ANTIOXIDANT IN INHIBITED THERMO-OXIDATION

As has been emphasized above for thermo-oxidation of many polymers, the intensification of colour as a result of the formation of polyconjugated regions (or PCS) outstrips or coincides with autobraking of the process. The data obtained allow us to state that 'self-stabilization' is connected with the inhibiting role of PCS.

On the other hand, we have come to the conclusion that monomer antioxidants are also able to form oligomers due to radical polyreactions under the conditions of inhibited thermo-oxidation.

The comparison of these facts allows the drawing of a conclusion about the possibility of an interaction between propagating the inhibitor chain with active centres of oxidizing polymer macromolecules. This process has been investigated by Kovarskaya, Livertovskaya and co-workers for the case of tetrahydrofuran–propylene oxide co-polymer thermo-oxidation inhibited by bisphenols[46].

In our laboratory this process has been investigated by Ivanov for atactic polypropylene oxidation ($T = 160°$ to $200°C$ and $P_{O_2} = 760$ mm Hg) inhibited by anthracene[4, 52].

The data obtained give rise to a conclusion about the formation of a thermodynamically more stable polyene conjugated system by atactic polypropylene thermo-oxidation at rather high temperature which may be partially aromaticized by thermal effects.

Taking all this into consideration, it is logical to assume that radicals formed in inhibitor oxidation may react with α-hydrogen of the CH_2 groups conjugated with polyene chains or with active centres which appeared as a result of the reaction with oxygen.

To elucidate the nature of the transformations, the product of polypropylene oxidation in the presence of 0.8 mol/kg of anthracene at 225°C

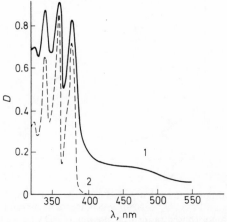

Figure 22. Absorption spectra of polypropylene–anthracene co-oxidation product (fraction S-1, curve 1) and anthracene (curve 2).

was accumulated within 100 min. The oxidized mixture was separated into two fractions soluble (S) and insoluble (IS) in cold heptane. Brown and dark brown benzene soluble (S-1) and insoluble (IS-1) fractions were obtained by the following more delicate fractionation by vacuum-distillation at 130° to 150°C (10^{-5} mm Hg) and treatment in boiling benzene. There are three absorption bands in the region 2840–2960 and 1460 cm^{-1} in the i.r. spectrum of S-1 which have been attributed to the CH_3 and CH_2 group absorption. Moreover, several bands corresponding to anthracene (1625, 1450, 1385, 959, 888, 727 cm^{-1}) have been observed. Anthraquinone bands have not been found. The spectra of both fractions have a considerable absorption background, the most intensive background absorption being observed for the benzene-insoluble fraction IS-1. The insoluble fraction IS-1 absorption bands are rather wide without narrow maxima. Nevertheless, one has a good basis to affirm the presence of aromatic polyconjugated structures (absorption at 750–900 and 1600 cm^{-1}), and also polyene blocks of conjugation formed from polypropylene and containing lateral CH_3 groups.

The electron absorption spectrum of the benzene-soluble fraction S-1 is shown in *Figure 22*. It contains absorption bands typical of anthracene (340, 360, 378 mμ) and a sufficiently intense long-wave background absorption without maxima.

Fractions S-1 and IS-1 have a narrow symmetrical e.p.r. signal (the PMC concentration is 6×10^{17} spin/g^{-1} for S-1 and 3×10^{18} spin/g^{-1} for IS-1).

The element analysis gives five per cent of oxygen in S-1.

The data obtained allow us to assume that S-1 and IS-1 are stable complexes of a polyconjugated polymer with anthracene and products of its transformation. This polymer is formed by thermo-oxidizing destruction of

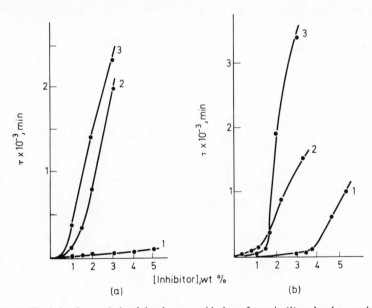

Figure 23. The induction periods of the thermo-oxidation of ceresin (1) and polypropylene (2) inhibited by anthracene: (a) at $T = 160°C$; (b) at $T = 180°C$.

201

Scheme V

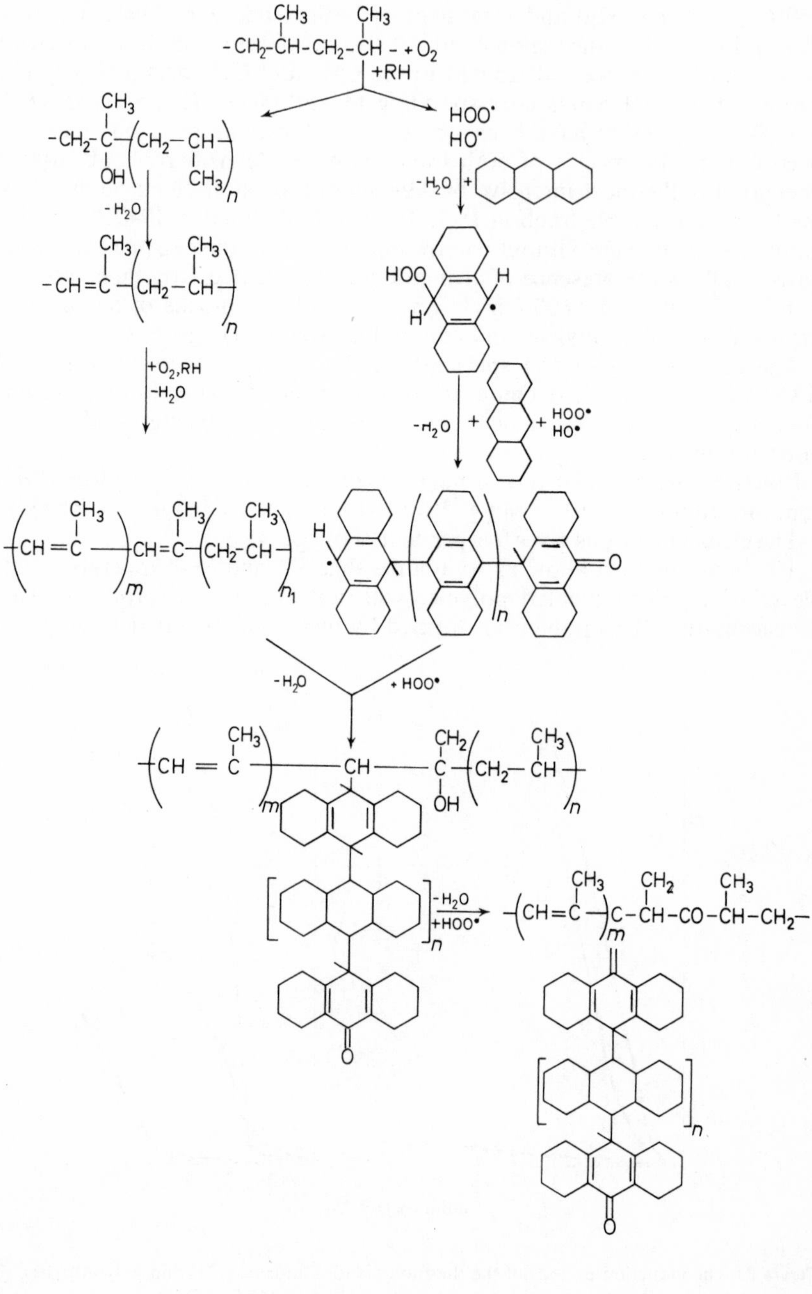

polypropylene in the presence of anthracene and anthraquinone and, apparently, is a graft copolymer of polyene with anthracene and polymer products forming in oxidizing dehydropolycondensation of anthracene containing PMC.

Significant inhibiting effects in polypropylene oxidation in the presence of anthracene are apparently connected with the interaction between the inhibitor and the oxidizing substrate when the reactions with oxygen, peroxide or alkoxy radicals take place (see *Figure 23*). The region of conjugation in polypropylene macromolecules already appears in the initial stage of oxidation. Anthracene molecules are taken into complexes with polyene conjugated blocks and after that their oxidizing dehydrocondensation takes place. The PSC paramagnetic fractions are accumulated and terminate the chain propagation and formation of hydroperoxides.

This process cannot be realized to the same degree in the oxidation of ceresin. The latter does not in practice contain tertiary hydrogen and so has considerably lower reactivity with respect to oxygen and peroxide radicals.

In a general form the hypothetical scheme may be represented as shown in *Scheme V*.

The investigation of the inhibiting activity of products as it appears in joint transformations of polypropylene and anthracene has shown that their activities are much higher than that of anthracene alone. It is interesting to note that these products are very effective in thermo-oxidation inhibition of both polypropylene and ceresin [*Figures 24*(a) and (b)] as well as other polymers.

It has been established that after an induction period of ceresin oxidation in the presence of the S-1 fraction the decrease in the absorption band inten-

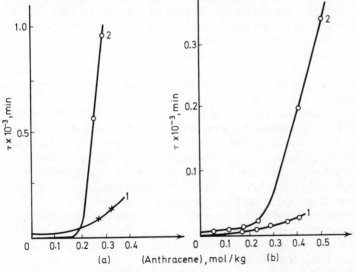

Figure 24. The inhibited thermo-oxidation of ceresin (a) and polypropylene (b) at 160°C in the presence of anthracene (1), polymer fraction of anthracene thermolysis (2), the fraction S-1 of the polypropylene–anthracene co-oxidation product (3)

sity of this fraction in the visible and u.v. region and PMC concentration considerably decreases. This is connected with the expenditure of anti-oxidant, with the decrease in the continuous conjugated chain in it as a result of the interaction with peroxy radicals. An analogous process of PMC decay has been observed at the end of the induction period for various PCS inhibitors.

CONCLUSIONS

Based on the data considered in the present report we come to the con-clusion that the theoretical predictions about the formation and the role of polyconjugated systems in thermolysis and thermo-oxidation of organic compounds are correct.

In practice, we have considered here the physical chemistry of the so-called 'collateral' reactions which determine to a considerable degree the kinetics of processes, and the structure and properties of the main sub-stances.

The data about these processes obtained at the present time must be considered the first stage of work leading to new and more powerful ideas about the destruction and also to the methods of polymer stabilization and inhibition of analogous processes for low-molecular weight compounds.

Undoubtedly, analogous processes proceed in photolysis and radiolysis, although the kinetics and chemistry of PCS formation and the influence of PMC and polyconjugation are virtually not investigated in these cases. It must be emphasized that the knowledge of the nature and structure of forming PCS and PMC, the information about the mechanism of their transformation caused by various energies and chemical effects, and especially the ability to inactivate radicals and to regulate the amount of PMC and PCS are of great importance for the problem of stabilization.

On the other hand, the understanding of these processes opens new ways for a purposeful preparation of highly effective polymer stabilizers. Appar-ently, an approach based on the uses of thermolysis, thermo- and photo-effects or ionizing irradiation of low-molecular weight antioxidants in the absence or in the presence of oxygen or some other oxidizers is rather auspicious.

The expediency of this approach has been illustrated by us by using PCS-stabilizers obtained by thermolysis or thermo-oxidation of anthracene, diphenylamine, and some other analogues as the example. The transition from InH-type inhibitors, to corresponding PCS with stable PMC is rather hopeful, because in this case it is possible to solve some complex problems of thermo-oxidation inhibition at a fairly high temperature ($T = 250°$ to $300°C$) and under the effect of u.v. or high energy rays.

Therefore, progress in the research on the effect of local activation and consequent intercombination transitions in the complexes of PMC or radicals stabilized by conjugation with the substances containing π-bond or heteroatoms is of particular importance.

It is also necessary because this effect may play a part in a great variety of processes with the participation of stable radicals.

One must bear in mind that the activation of transition and reactivity may be realized not only by the action of polyconjugated PMC but also due to such stable radicals as O_2, NO, etc.

These predictions may be confirmed by some examples, and in particular by the influence of oxygen and nitric oxide on the rate of CCT formation[28] where for the first time the catalytic role of oxygen in thermodestruction of aromatic polyamides has been shown by means of ^{18}O (ref. 26).

All this gives basis for the conclusion that progress in work on the effect of local activation allows us to see many chemical problems including the transformation of low-molecular and polymer substances due to energy effects and oxidation.

REFERENCES

1 A. Wall, *Soc. Plast. Engrs J.* No. 8, 811 (1960).
2 A. A. Berlin, *J. Macromol. Sci.-Chem.* **A5**, 1187 (1971); *Vysokomol. Soedin.* **A13**, 2429 (1971).
3 A. A. Berlin, *Izvest. Akad. Nauk SSSR, Ser. Khim.* 59 (1965); *Khim. Prom.* **12**, 23 (1962); *Izvest. Akad. Nauk SSSR, Otdel. Khim. Nauk*, 1494 (1962); *Dokl. Akad. Nauk SSSR*, **150**, 795 (1963).
4 A. A. Berlin, 'Kinetics and mechanism of polyreactions', *International Symposium on Macromolecular Chemistry*, IUPAC: Budapest (1969), Plenary and Main Lectures, p 689; *Vysokomol. Soedin.* **A13**, 276 (1971).
5 A. A. Berlin, L. A. Blyumenfeld and N. N. Semenov, *Izvest. Akad. Nauk SSSR, Otdel. Khim. Nauk*, 1689 (1959).
6 A. A. Berlin, B. E. Davydov and B. Krentsel *et al. Chemistry of Polyconjugated Systems*, p 223. Khimiya: Moscow (1972).
7 B. A. Dolgoplosk, S. J. Beilin and J. N. Marevitch, *Vysokomol. Soedin.* **A13**, 1278 (1971).
8 A. A. Berlin, *Chem. Technol. Polymers*, No. 7–8, 134 (1960); *Chem. Prom.* No. 5, 375, 6, 144 (1960).
9 A. A. Berlin, *Preprints* IUPAC International Symposium on Macromolecular Chemistry, Prague, p 281 (1965).
10 A. A. Berlin, G. A. Vinogradov and A. A. Ovchinnikov, *Internat. J. Quantum Chem.* **6**, 263 (1972).
11 V. A. Benderskii, L. A. Blyumenfeld and D. A. Popov, *Zh. Strukt. Khim.* **7**, 370 (1966).
12 L. A. Blyumenfeld, V. A. Benderskii and P. A. Stundzhas, *Zh. Strukt. Khim.* **7**, 686 (1966).
13 Kh. M. Gafurov, *Thesis*, Akademija Nauk Uzbekskoj SSR: Tashkent (1967).
14 I. D. Morosova and M. Je. Dyatkina, *Uspekhi Khim.* **37**, 865 (1968).
15 A. A. Berlin, V. A. Grigorovskaya, V. P. Parini and Kh. Gafurov, *Dokl. Akad. Nauk SSSR*, **156**, 1371 (1964).
16 A. A. Berlin, V. A. Grigorovskaya, V. K. Skachkova and V. E. Skurat, *Vysokomol. Soedin.* **A10**, 1578 (1968).
17 A. A. Berlin and N. G. Matveeva, *Vysokomol. Soedin.* **8**, 736 (1966).
18 A. A. Berlin, M. I. Cherkashin and V. G. Zadontsev, *Izvest. Akad. Nauk SSSR, Ser. Khim.* 2065 (1967); *Vysokomol. Soedin.* **B9**, 91 (1960).
19 A. A. Berlin, A. P. Firsov and V. V. Yarkina, *Vysokomol. Soedin.* **B10**, 724 (1968).
20 A. A. Berlin, M. I. Cherkashin, M. V. Borshilova, G. Bantsirev and J. M. Panaiotov, *Vysokomol. Soedin.* **A14**, 9 (1972); *International Conference on Chemical Transformations of Polymers, IUPAC*, Vol. 2, p 38, Bratislava (1971).
21 A. A. Berlin, V. P. Parini and K. Almanbetov, *Dokl. Akad. Nauk SSSR*, **166**, 595 (1966).
22 A. A. Berlin, R. N. Belova and A. P. Firsov, *Dokl. Akad. Nauk SSSR*, **180**, 140 (1968); *Vysokomol. Soedin.* **B10**, 366 (1968).
23 A. A. Berlin and V. M. Kobryanskii, *Izvest. Akad. Nauk SSSR, Ser. Khim.* (in press).
24 A. A. Berlin, Kh. M. Gafurov, V. F. Gakhkovsky and V. P. Parini, *Izvest. Akad. Nauk SSSR, Ser. Khim.* 728, 746 (1966).
25 Kh. M. Gafurov, V. M. Mulikov, V. F. Gakhkovsky, V. P. Parini, A. A. Berlin and L. A. Blyumenfeld, *J. Struct. Chem.* **6**, 649 (1945).
26 A. B. Blyumenfeld, N. Ya. Valetskaya, B. M. Kovarskaya and A. S. Akutin, *Vysokomol. Soedin.* **B14**, 403 (1972).

[27] A. L. Buchachenko, *Thesis*, Institute of Chemical Physics, Akademija Nauk SSSR; Moscow (1968).

[28] S. B. Goldstein, S. D. Stavrova, J. P. Saveleva and S. S. Medvedev, *Dokl. Akad. Nauk SSSR*, **181**, 134 (1968).

[29] M. V. Bazhilova, *Thesis*, Institute of Chemical Physics, Akademija Nauk SSSR: Moscow (1972).

[30] A. A. Berlin, G. A. Vinogradov and V. M. Bobryanskii, *Izvest. Akad. Nauk SSSR, Ser. Khim.* 1192 (1970).

[31] N. J. Chernorukov and S. E. Krein, *Oxidation of Mineral Oils*, Gostopisdat: Moscow (1946).

[32] S. R. Sergienko and N. J. Cherniak, *Dokl. Akad. Nauk SSSR*, **113**, 331 (1957)

[33] G. E. Zaikov, E. A. Blumberg and N. M. Emanuel, *Neftekhim.* **5**, No. 1, 53 (1965).

[34] A. V. Bobolev, E. A. Blumberg, E. G. Rozantsev and N. M. Emanuel, *Dokl. Akad. Nauk SSSR*, **80**, 1139 (1968).

[35] A. A. Berlin and G. V. Belova, *Uspekhi Khimii i Fizicheskoi Khimii Polimerov*, p 3. Nauka: Moscow (1970).

[36] A. A. Berlin, *Uspekhi Khim.* (in press).

[37] A. A. Berlin, A. P. Firsov and V. V. Yarkina, *International Symposium on Chemical Transformations of Polymers, IUPAC*, Vol. 2, p 34: Bratislava (1971).

[38] V. V. Yarkina, *Thesis*, Moscow State University: Moscow (1968).

[39] V. K. Belyakov, A. A. Berlin, V. P. Dubyana, L. V. Nevski and O. G. Tarakanov, *Plastmassy*, No. 8, 35 (1968); *Vysokomol. Soedin.* **A8**, 1677 (1967).

[40] A. A. Berlin, V. K. Belyakov, V. A. Orlov and O. G. Tarakanov, *Kinetika i Kataliz*, **3**, 687 (1967).

[41] B. K. Belyakov, J. M. Bukin, O. G. Tarakanov and A. A. Berlin, *Vysokomol. Soedin.* **A13**, 1585 (1971).

[42] A. A. Berlin, B. E. Davydov, B. Krentsel *et al. Chemistry of Polyconjugated Systems*, p 72: Khimiya: Moscow (1972); *Vysokomol. Soedin.* (in press).

[43] A. A. Berlin and S. I. Bass, *Starenie i Stabilizatsya Polimerov*, Khimiya: Moscow (1966).

[44] A. A. Berlin, A. A. Ivanov and A. P. Firsov, *Vysokomol. Soedin.* **B12**, 80 (1970).

[45] L. L. Yasina, B. A. Gromov, V. B. Miller and J. A. Shlyaptsinov. *Vysokomol. Soedin*, **8**, 1411 (1966).

[46] J. J. Levantovskaya, V. V. Guryanova, B. M. Kovarskaya and J. J. Slonim, *Vysokomol. Soedin.* **A11**, 1043 (1969).

[47] A. Ivanov, *Thesis*, Moscow State University: Moscow (1969).

[48] A. A. Berlin, A. A. Ivanov, J. J. Mirotvortsev and G. K. Goryacheva, *Vysokomol. Soedin.* **B12**, 563 (1970). *XIth Microsymposium on the Mechanism of Inhibition Processes in Polymers, IUPAC*, Prague (1972).

[49] J. Gömöry, A. Gömöryova, R. Reya and J. Stimel, *Europ. Poly. J.—Suppl.* 545 (1969); **6**, 1047 (1970).

[50] J. Gömory, J. Stimel, R. Reya and A. Gömöryova, *J. Polymer Sci.* **C16**, 451 (1967).

[51] J. Pospíšil, E. Lisá and J. Buden, *Europ. Polym. J.* **6**, 1347 (1970).

[52] A. A. Berlin, A. A. Ivanov and V. J. Popovkina, *Vysokomol. Soedin.* **A13**, 2724 (1971).

TRANSFORMATIONS OF PHENOLIC ANTIOXIDANTS DURING THE INHIBITED OXIDATION OF POLYMERS

J. Pospíšil

Institute of Macromolecular Chemistry, Czechoslovak Academy of Sciences, Prague, Czechoslovakia

ABSTRACT

During ageing and oxidative degradation of stabilized polymers, transformations of antioxidants take place as a result of reactions with radicals RO$^{\cdot}$ and ROO$^{\cdot}$, alkyl hydroperoxides and/or oxygen. Knowledge of these processes as well as of the properties of products thus formed is necessary for a complex evaluation of antioxidants. Data are given dealing with typical products of oxidative transformations of phenolic antioxidants prepared by independent syntheses, with model transformations under conditions simulating interactions in a stabilized organic substrate, with products identified under real conditions of oxidation of a stabilized substrate, and with influence of some transformation products on the course of oxidation of tetralin and polypropylene.

The oxidative radical degradation of polymers which takes place under most varied accompanying conditions during the atmospheric ageing or processing of polymers is limited by stabilization. The importance of the stabilization process and the theoretical interest devoted to it are an unequivocal consequence of the technical and commercial interests. The majority of the antioxidants technically used have an empirical basis. Complex requirements dealing with the properties of the stabilization systems have been gradually formulated. The data acquired so far about the stabilization processes and the chemical structure of the stabilizers have been summarized in a number of monographs[1-7]. Every year suggestions are also made concerning the use of new stabilization systems.

When choosing the antioxidants, one concentrates on attaining the maximum activity while preserving at the same time the optimum technical properties of the polymer. In the initiation stage of the degradation process, free radicals are formed from the oxidized polymer, RH

$$RH \xrightarrow[\text{X}^{\cdot}]{\text{initiation}} R^{\cdot} \tag{1}$$

According to the generally accepted scheme of autoxidation, alkyl peroxyls and alkyl hydroperoxides are formed in the propagation step of the reaction; they are a source of further radicals which play their role in the degradation process

$$R^{\cdot} + O_2 \rightarrow ROO^{\cdot} \tag{2}$$

$$ROO^{\cdot} + RH \rightarrow ROOH \tag{3}$$

207

The stabilization of polymers against the oxidative degradation is based on a preventive protection against radical formation, on the deactivation of the peroxyls already formed, and on the non-radical decomposition of hydroperoxides. I shall consider here only compounds breaking the kinetic chain by their reaction with ROO˙. These compounds need not necessarily be free radicals themselves. However, they must be able to react comparatively readily with ROO˙ while giving rise to other radicals, so stable that they cannot initiate new kinetic chains. For this purpose, it has become a tradition to use a group of oxidation inhibitors (InH) containing a labilized hydrogen atom in their molecule. From the viewpoint of the polyolefins, phenolic compounds are those most important of all. The fundamentals about the relationships existing between the structure and the antioxidative activity have been defined[8], and any new development of the required construction of the molecule is already greatly restricted. No extraordinary novel results as to the antioxidative activity can be expected—the activity is determined by the mechanism of their action. However, the development leads to some partial modifications of the molecule of a phenolic antioxidant, to combinations with fragments containing heteroatoms and introducing into the molecule the ability of a polyfunctional action. Above all, these modifications provide activity at high temperatures of the oxidative degradation, when the phenols alone are quickly deactivated and therefore completely useless. For these compounds, the greatest potential development of types can be expected. They can be fully used up only under special conditions and at the processing temperatures. However, for the greatest part of the time of its practical applicability, the polymer is exposed to atmospheric temperatures, when just the monofunctional phenolic antioxidant is very useful. We possess a wide scale of information about the technical properties and toxicity of such compounds. For practical reasons, the terms discolouring and staining antioxidant have been introduced. It is just this property of the phenolic antioxidants, as well as a general survey of the transformations during the process of the inhibited oxidation leading to the consumption of the antioxidant, which are the subject of this lecture.

A phenolic antioxidant (InH) acts by breaking the kinetic chain of the autoxidation

$$InH + ROO˙ \rightarrow In˙ + ROOH \qquad (4)$$

During the process, a radical In˙, which can be primarily regarded as an aryloxyl, and a hydroperoxide are formed. An analogous aryloxyl arises by a reaction with alkoxyls RO˙

$$InH + RO˙ \rightarrow In˙ + ROH \qquad (5)$$

which are formed from alkyl peroxyls or alkyl hydroperoxides arising during the autoxidation. The aryloxyl In˙ is transformed into more stable products. From the viewpoint of an exact knowledge of the stabilization process, we are interested to know the structure of the products formed under the specific conditions of atmospheric ageing and thermo-oxidative degradation, and also the properties of the transformation products.

I shall first give the factors which can generally participate in the deactivation and transformation of an antioxidant in the process of inhibited oxidation during atmospheric ageing and thermo-oxidative degradation of polymers:

(a) Active radical species formed during autoxidation from the substrate RH (R˙, ROO˙, RO˙, HO˙) and the alkyl hydroperoxide ROOH appearing as the primary oxidation product in the chain reaction. A certain role in the process of deactivation can also be played by the secondary oxidation products of substrate (in the first place, compounds containing the hydroxyl and the carboxyl group) and by accumulating reactive transformation products of the antioxidant, aryloxyls in the first place.

(b) Air oxygen or traces of ozone and nitrogen oxides present in the air.

(c) The trace amounts of catalytically active impurities in the polymer, including the residues of the polymerization catalyst.

(d) Temperature of the process of the co-effect of radiation.

A mere general enumeration of factors points out the complicated character of the processes to be expected during the transformations. Besides chemical reactions and in spite of the low concentration of the antioxidant in the polymer there emerge possibilities of physical interactions causing changes in reactivity. These relationships are completely beyond the framework of this investigation.

It can be said, therefore, that an antioxidant of the InH type capable of breaking the kinetic autoxidation chains is transformed during the process, its original active form being consumed. It has been pointed out that at the end of the induction period the antioxidant added to the mixture is already mostly consumed by the inhibition reaction (active consumption), or that its concentration has fallen off to such a degree that it is no longer able to interfere in an active way with the propagation step of the autoxidation. As a rule, the consumption of the antioxidant in this step of the process is schematically represented by

$$In˙ + ROO˙ \rightarrow InOOR \tag{6}$$

$$2In˙ \rightarrow In—In \tag{7}$$

the overall process is often described as a formation of 'inactive products' from the antioxidant. As will be seen further, in no case is it possible to agree with this designation, as long as we do not know more exactly their character and the conditions of the degradation process. Thus, a complex characterization of the antioxidant under investigation requires—besides the knowledge of its activity—also the knowledge of the products arising therefrom during inhibited oxidation. An investigation of the formation of the transformation products of the antioxidants directly in the stabilized substrate is very difficult and must be preceded by a detailed study which can generally be divided into four stages:

(1) Obtaining a survey of the possible intermediates and oxidation products of the antioxidant under study. For this purpose, independent syntheses involving reagents leading to good yields of the individual well-defined products are used advantageously. The processes optimal from both the preparative and mechanistic viewpoints are chosen here, although they fre-

quently differ from the processes which can be considered during inhibited oxidation. The products thus obtained are used as identification standards for the following model investigations, as well as for obtaining the necessary data about the chemical and physical properties under specific experimental conditions. The state of knowledge existing so far and the classification of information on the oxidative transformations of phenols allow a comparatively easy orientation, at least as to the main features of the problem studied here.

(2) Investigation of a model simulated reaction between an antioxidant and the $ROO^{\cdot}$ or $RO^{\cdot}$ radicals and a reaction with hydroperoxide and oxygen. It is only with great difficulty that one can perceive all reactions occurring during ageing in a real system. The results of a model investigation involving reactions with the above mentioned agents can give on the whole a real picture of the behaviour of an antioxidant during thermal oxidation. At a sufficient partial pressure of oxygen, the antioxidant terminates the kinetic chains by an interaction with alkylperoxyl. An interaction with the $R^{\cdot}$ radicals need not be considered in a common case. However, such reaction can be taken into account in some special cases of the polymer oxidation, e.g. in the case of an insufficient access of the air oxygen. To identify the products of model reactions, standard compounds prepared as indicated sub (1) are used.

The model process more or less simulates interactions during the inhibited oxidation itself. The same partners take part in the model reaction, but it is carried out under different concentration, and very often also temperature, conditions. Although the results thus obtained are very illustrative for the transformations themselves and permit numerous conclusions about the course of the individual reactions, they nevertheless cannot be applied to the stabilization process proper without verifying them in a real inhibitive system.

(3) Oxidation of the stabilized substrates under conditions allowing an investigation of the changes in the structure of the antioxidant in various stages of the processes and under various conditions of oxidation. Preparation of products is extremely difficult owing to the necessity of isolating them from dilute solutions; as a rule, it is successful only for more stable compounds. The identification of the radical species is usually impossible; the mechanism of transformations can be considered only on the basis of knowledge of real analogies with the mechanisms of model investigations and of knowledge of the reactivity of the intermediates or primary reaction products.

(4) Determination of the effect of the transformation products of the antioxidant under investigation on the oxidized substrate under conditions for which their formation has been proved. The effect of the individual compounds and the mixtures thereof, or mixtures with the original antioxidant, must be known. Owing to the thermal sensitivity of some products or to their easy volatility, the effects on the autoxidation process should be followed even under mild conditions. Oxidation of low-molecular weight hydrocarbons instead of polymers, tetralin in the first place, has proved to be useful. The results can be extrapolated with a comparatively good accuracy also to the conditions of the oxidation of polyolefins.

The consideration concerning the formation of the transformation products of monohydric phenolic antioxidants allows us to take into account the formation and behaviour of the aryloxyl In˙ which can arise during the inhibition process by reactions with oxyradicals 4 and 5 or by analogies thereof, by a reaction with the carbon radical R˙

$$InH + R˙ \rightarrow In˙ + RH \qquad (8)$$

or by oxidation with oxygen

$$InH + O_2 \rightarrow In˙ + HOO˙ \qquad (9)$$

Neither is it possible to exclude the formation of an aryloxyl due to radiation

$$InH \rightarrow In˙ + H˙ \qquad (10)$$

As to the reactions of the aryloxyl In˙, we are interested in the first place in the products of oxidative coupling 7 which models the transformations of the antioxidant occurring without participation of alkoxyls or alkyl peroxyls, and the transformation products of reaction 6 or of its analogy with the RO˙ or HOO˙ radicals. With respect to the structure of the antioxidants and to the reactivity of the aryloxyls derived therefrom, we do not assume a reaction with the saturated hydrocarbon substrate RH

$$In˙ + RH \rightarrow InH + R˙ \qquad (11)$$

However, the formation of addition products with hydrocarbons containing an activated C=C double bond cannot be excluded[9].

The generation and reactions of aryloxyls derived from monohydric phenols have been described in the literature[10, 11]. The aryloxyls react in their mesomeric forms as oxy- or carbon radicals. The charged mesomeric structures also contribute to their physical properties. For the aryloxyl to exist in a more stable form than a mere transient state during the reaction, its positions ortho and para must be substituted by groups causing resonance and steric stabilization. It is the same structural requirement which we know from the investigation of the relationships existing between the structure and activity of the phenolic antioxidants[8]. Aryloxyls are then able to exist also in the form of polyradicals[12]. However, they were not identified during the inhibited oxidation of polymers and their formation can be deduced only from the presence of more stable products.

The rate at which the primarily and transiently formed aryloxyls are transformed into further products is affected by their stability, by the momentary concentration of aryloxyls as well as radicals ROO˙, and by reactivity of the surrounding medium. At different temperatures of inhibited oxidation, essential changes in the mechanism of their reactions must also be expected. Let us consider here typical reactions of the mono- and polynuclear aryloxyls derived from monohydric phenolic antioxidants which must be taken into account in the discussion of changes during ageing and thermo-oxidative degradation of polymers.

Oxidative coupling according to the general scheme 7 can lead in the case of mononuclear aryloxyls to the formation of new C—C and C—O bonds. The joining of two aryloxyls by the O—O bond is fully excluded. The reaction can occur repeatedly, with a mixture of primary and consequent products having the character ranging from dimers to polymers as the result. The total result of the reaction is influenced by the structure of the original oxidized phenol. Simpler mixtures of products arise from sterically hindered phenols. The extent of the dimerization is determined by mesomeric and especially steric effects. The latter can completely suppress the dimerization[13]. Most of the sterically hindered aryloxyls undergo dimerization C—O. Quinolethers Ia, Ib are formed by this reaction. Such products arise from 2,4,6-trialkylphenols, 2,6-dialkyl-4-alkoxyphenols, or from 2,4-dialkyl-6-alkoxyphenols, e.g. refs. 14–16.

The position of the C—O bond formed is affected by the volume of the substituents in the position ortho or para to the phenolic hydroxyl. The steric factor also influences the equilibrium between aryloxyl and its dimer in solution; bulky substituents shift the equilibrium to aryloxyl.

From 2,6-dialkylphenols, products ranging as far as polymers are formed by route of a mechanism involving the disproportionation of quinolethers[19,20]. Intramolecular and intermolecular reactions of aryloxyls give rise to a mixture of products during the transformation of 1,1'-dinaphthyl-2,2'-diol[21].

The dimerization by means of a C—C bond occurring to positions 2 and 4 is typical for the di- and trialkylated phenols. Several cases have been described[10, 14]; cyclohexadienoid (II) or diphenoquinoid (III) compounds are formed as the resulting products. In II, R^3 = H, a rapid rearrangement to bisphenol takes place. The dimerization by the C—C bond and the formation of a further intramolecular bond can occur during the oxidation of 2,4-dialkylphenols[17]. Similar changes were also observed for 2-tert-butyl-4-methoxyphenol[18], an important antioxidant. A spiro compound IV is formed.

In the case of an appropriate substitution of the ring, even a rearrangement of the radical can occur during the oxidative coupling of the mesomeric carbon radicals. Such a mechanism including formation of a benzyl radical was considered for the oxidation of 2,6-di-tert-butyl-4-methylphenol where a phenolic (V) and a quinoid (VI) dimer are formed[22]. However, there are no spectral proofs for the formation of the benzyl radical. A chemical proof of its transient formation in the course of the oxidation of 2,6-di-tert-butyl-4-methylphenol has been provided by the isolation of 2,6-di-tert-butyl-4-(3,5-di-tert-butyl-4-hydroxybenzyl)-4-methyl-2,5-cyclohexadienone[23](VII).

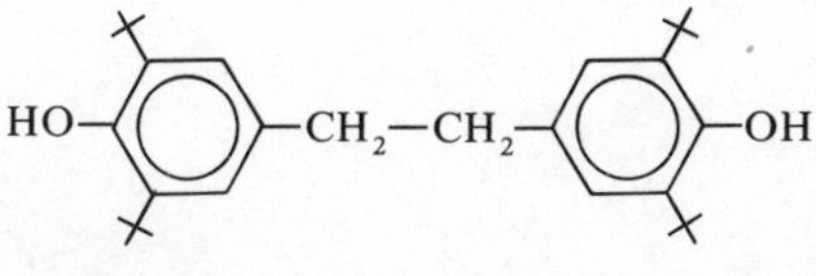

V

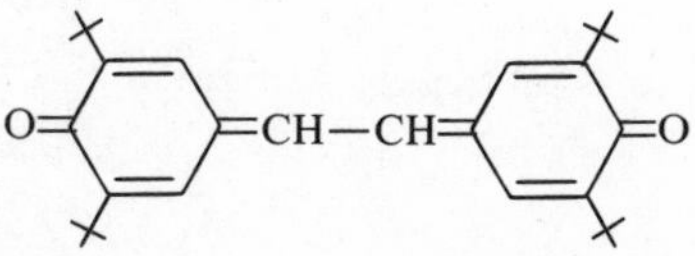

VI

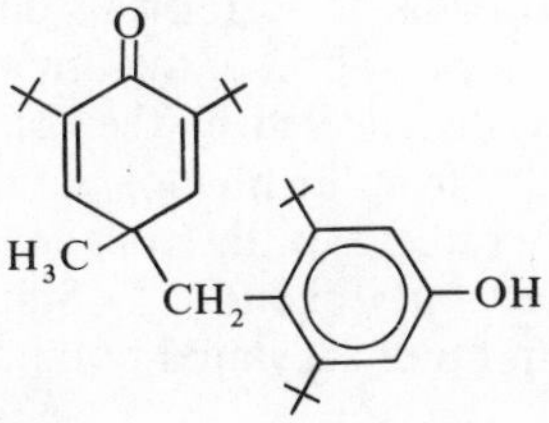

VII

Aryloxyls derived from 2,4,6-trialkylphenols having α-hydrogen in their para-substituent undergo rather disproportionation than dimerization. This is accompanied by the formation of the original phenol and quinone methide, e.g. VIII, which is transformed further according to its stability[24, 25].

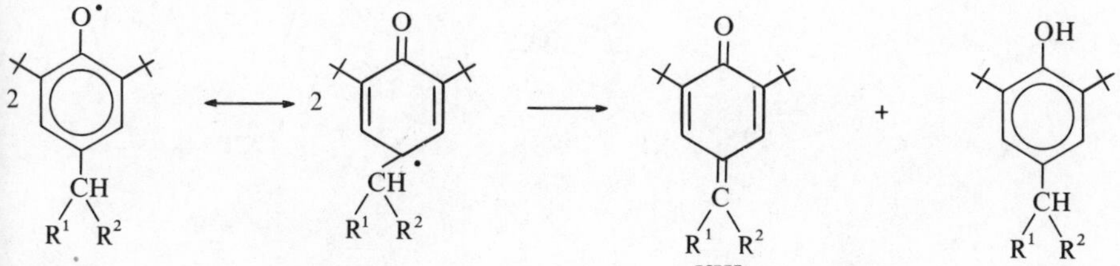

VIII

Quinone methide VIII having R^1, $R^2 = H$ is very unstable and cannot be isolated. On this basis, the formation of compounds V and VI as oxidation products of 2,6-di-tert-butyl-4-methylphenol is interpreted[26]. (An alternative mechanism for the above dimerization of two benzyl radicals.) The stability

213

of quinone methide is enhanced by the substitution of some of the α-hydrogens in the para substituent with an alkyl. For instance, quinone methide derived from 2,6-disubstituted para-isopropylphenol is stable for several days.

For phenol derivatives containing a methylene group at the position para, a slow oxidation to the carbonyl group[27] and the formation of a derivative of acetophenone IX can be expected.

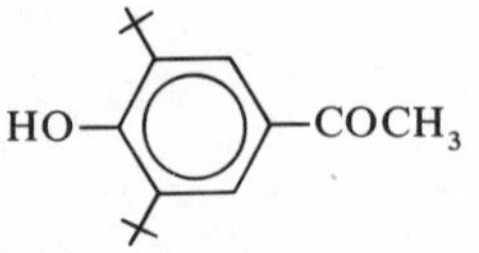

IX

Aryloxyls are also formed from polynuclear phenols which belong to the technically important antioxidants. In radicals derived from the alkylidene- and thiobisphenols, the delocalization of the unpaired electron in the system of both six-membered rings is affected by the character of the bridge. More stable radicals are formed[28] from bisphenols having the bridge at positions 4,4'. One of the most stable radicals is the so-called galvinoxyl X derived from 4,4'-methylenebis(2,6-di-tert-butylphenol)[29]. Similar stable radicals, e.g. XI or XII, are formed from other alkylated polynuclear phenols[30, 31].

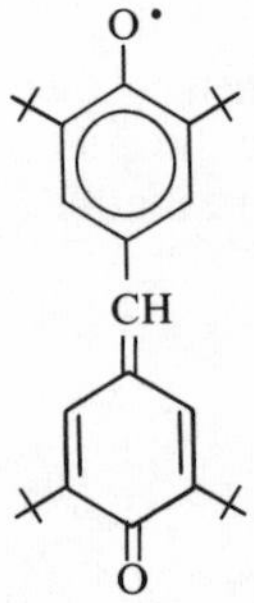

X 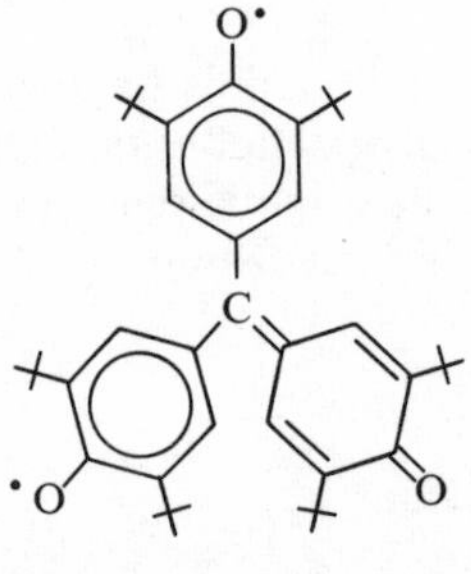   XI 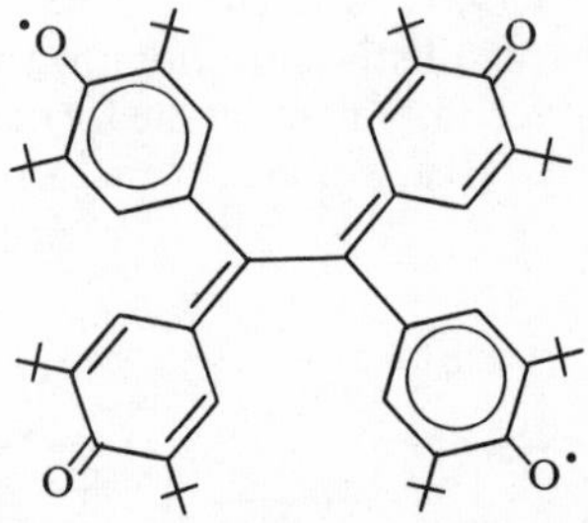   XII

Such stable radicals can even participate in a transient way in the discolouration of the stabilized substrate during atmospheric ageing. In some cases, the aryloxyls derived from 4,4'-alkylidenebisphenols can undergo an intramolecular coupling with the formation of spirobisdienone. For instance, 4,4'-isopropylidene-bis-(2,6-di-tert-butylphenol)[32], which belongs to the group of nonstaining, but little active antioxidants, is oxidized in this way.

The cyclohexadienone derivative thus obtained has the structure XIII. A transformation of another antioxidant of similar structure having ortho-methyl-substituents which improve its antioxidative activity, can lead[33] to a complex polymeric product (XIV).

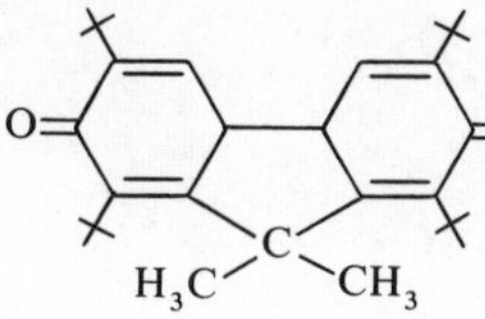

XIII

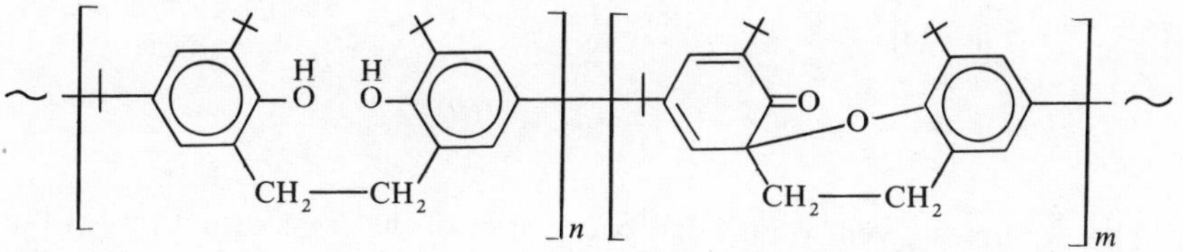

XIV

Oxidation of 4,4'-diphenyldiols gives rise to a biradical only transiently; diphenoquinones[34] (e.g. XV) are formed as the product.

From 2,2'-alkylidenebisphenols which belong to the group of important antioxidants, thermally stable cyclic quinolethers[35, 36] (XVI), can be formed in the case of an appropriate substitution of the aromatic ring which will not allow an intermolecular reaction.

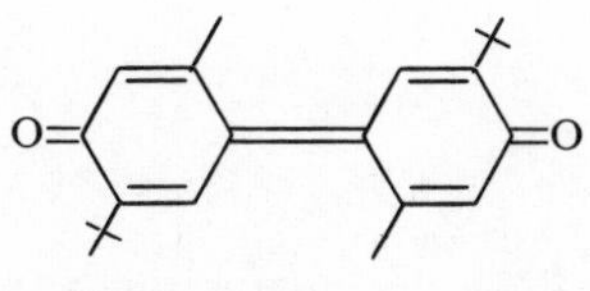

XV

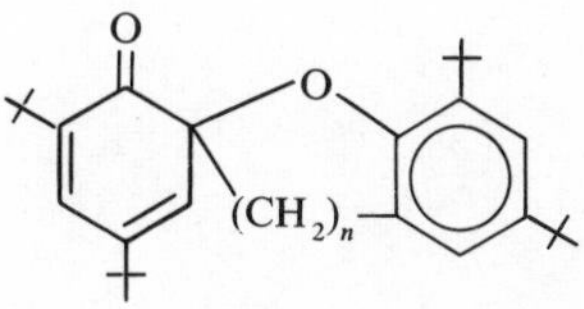

XVI

During the oxidation of binuclear phenols having their rings bonded by oxygen, sulphur or selenium atoms at positions 4,4', and tert-butylated at the positions ortho to the hydroxyl, the heteroatom is eliminated in the course of the oxidation reaction. 3,3',5,5'-Tetra-tert-butyl-4,4'-diphenoquinone (IV, R^1, R^2 = tert-butyl)[37, 38] appears here as the product. The oxidation of the sterically less hindered diphenyl ether proceeds with the formation of a more complex product[38].

215

The examples given so far have been those of the transformations of the individual antioxidants. A more complicated interaction can occur in the mixture of phenols, giving rise—in the case of mononuclear phenols—to unsymmetrical cyclohexadienone ethers[39-41]. The structure of the product is determined by steric factors; the sterically more hindered residue is bonded in the cyclohexadienoid part of the molecule, e.g.

A number of such products have been described. The interaction between the sterically hindered and unhindered phenols and mixtures of aryloxyls and more stable products derived therefrom can serve as an explanation of the homosynergism of the phenolic antioxidants[42].

The process of inhibited oxidation can lead not only to the transformation of the phenolic antioxidants caused by their reaction with the ROO• radicals or with oxygen, but also—given adequate conditions—to products in which both these agents are chemically bonded. The reaction with alkyl peroxyls is a termination one, and is generally schematically represented by 6. Here, the aryloxyl reacts in one of its mesomeric forms, according to the steric effect of the substituents on the ring. The stability of products thus formed is affected by their structure; some peroxy derivatives do not decompose until at temperatures above 100°C, so that their formation during the ageing of the polymers is quite likely. The reaction products most frequently studied were those of the reaction with tert-butylperoxyl, which can be regarded as the simplest alkylperoxyl modelling tertiary hydroperoxyls arising during the autoxidation of polypropylene. A number of reactions have also been carried out with cumylhydroperoxyls or tetralylhydroperoxyls under conditions suitable for the modelling of the transformation process during inhibited oxidation. The ROO˙ radicals needed for the reaction are usually generated by a homolytic decomposition of hydroperoxide catalysed with the Co˙˙ ions[43, 44] or by the thermal decomposition of the readily splitting hydrocarbons or azo compounds in the presence of oxygen[43]. In both cases of the generation, the RO˙ radicals are also contained in the mixture. Their presence can be excluded by working in an excess of hydroperoxide, when the reaction

$$RO˙ + ROOH \rightarrow ROH + ROO˙ \tag{12}$$

takes place. To investigate the reaction of the antioxidant with the RO˙ radicals (mostly tert-butoxyls), the necessary radicals can be unequivocally generated by the thermal decomposition of tert-butyl peroxide[45] or tert-butyl peroxalate[46].

I shall give here some typical products formed from the individual main types of phenolic antioxidants by their reaction with tert-butylperoxyl. 2,6-Dialkylphenols yield[47] 2,6-dialkyl-1,4-benzoquinone and diphenoquinone III. A varied mixture of products is obtained from 2,4-di-tert-butylphenol[48]. Apart from C—C and C—O dimerization products whose structures have been mentioned before, compounds XVII and XVIII are given as the examples of isolated peroxycyclohexadienones.

XVII   XVIII

Antioxidants having the structure of 2,4,6-trialkylphenols form 4-alkylperoxy-2,5-cyclohexadienones XIX or 2-alkylperoxy-3,5-cyclohexadienones (XX). The product formed from one of the phenolic mononuclear antioxidants most frequently used, namely, 2,6-di-tert-butyl-4-methylphenol[43, 49], belongs to the first type. It is reported[50] that thermal decomposition of alkylperoxycyclohexadienone XIX leads to 2,6-dialkyl-4-hydroxybenzaldehyde. Isomeric cyclohexadienones of type XX are formed from phenols containing a less bulky substituent R^2, e.g. from 2,4-di-tert-butyl-6-methylphenol.

217

XIX   XX

Compounds analogous to XIX and XX (R = H) are obtained by the autoxidation of trialkylphenols in an alkaline medium. For instance, from 2,4,6-tri-tert-butylphenol we obtain a mixture of the 2- and 4-cyclohexadien-one hydroperoxides[51], while only the 4-isomer is formed from 2,6-di-tert-butyl-4-methylphenol[52].

The isolation of alkylperoxycyclohexadienones XIX or XX in the reaction of alkylperoxyl with mononuclear phenolic antioxidants is in accordance with the assumption that a mononuclear phenol is able to terminate two kinetic oxidation chains, which is expressed in a general form by reactions 4 and 6. This is a chemical contribution to the proof of the mechanism of the effect of the antioxidants breaking the autoxidation chains. Kinetic proof, and particularly the rate-determining step, have been for a long time an object of discussions. However, this problem also seems to have been solved satisfactorily by applying suitable experimental techniques.

Some products of the transformations of polynuclear antioxidants with the ROO˙ radicals have also been described. In our laboratory, alkylidene-bisphenols have been studied in more detail. It has been found that some of the most active of them are transformed in polypropylene into deeply col-oured products. In the series of 2,2'-alkylidenebisphenols, more active antioxidants were represented by compounds having hydrogen atoms on the carbon atom of the bridge and substituted in the benzene rings by a bulky alkyl at the position ortho to the hydroxyl and by the methyl group at the position para. Substitution of all hydrogens on the bridge with alkyls or replacement of methyl with the tertiary butyl group reduced the discoloura-tion while at the same time also reducing the antioxidative activity. We have investigated in more detail the transformation products of a technically important antioxidant, namely, 2,2'-methylene-bis-(4-methyl-6-tert-butyl-phenol) which belongs to the most active alkylidenebisphenols for the stabili-zation of isotactic polypropylene. Although it is classified among the non-staining antioxidants, however, it substantially discolours polyolefins. It has been in use for a long time now, but nothing has been known about its transformation products. By using the reaction with tert-butylperoxyls, we prepared[53,54] several products whose formation depends on the molar ratio of the reacting components. In a mixture in which the tert-butyl hydro-peroxide used did not attain the equimolar ratio with respect to 2,2'-methyl-ene-bis-(4-methyl-6-tert-butylphenol), dimer XXI, trimer XXII, and com-pounds having an intensive brown colour were formed as the main products.

Their amount increased with the increasing hydroperoxide content in the mixture. Compounds XXI and XXII were also obtained by the reaction of 2,2'-methylenebis(4-methyl-6-tert-butylphenol) with the tert-butoxyl radicals.

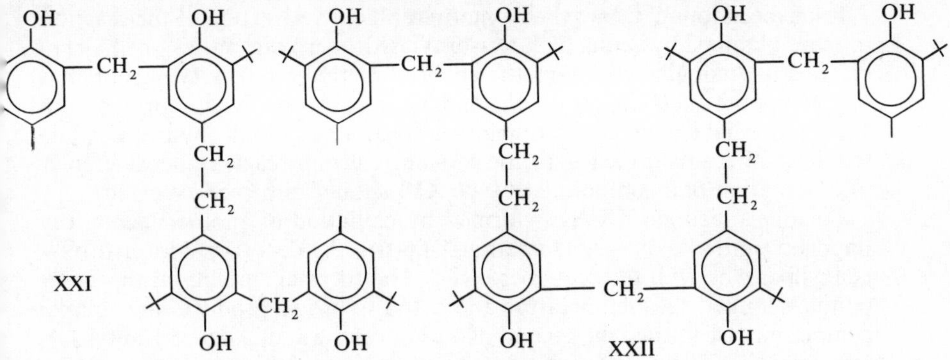

XXI

XXII

A large excess of tert-butyl hydroperoxide in the reaction mixture gave rise to the cyclohexadienone derivative XXIII and its isomer XXIV.

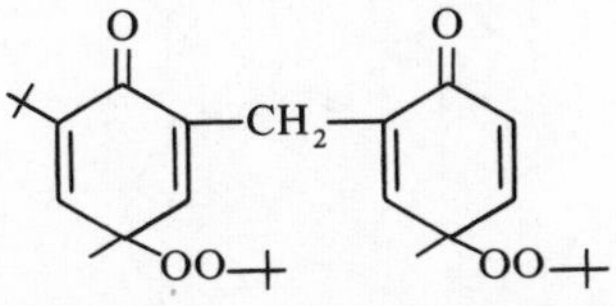

XXIII

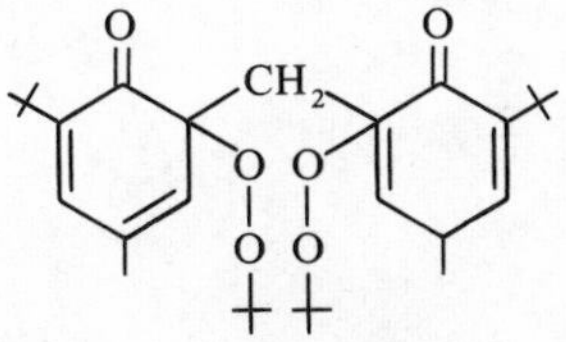

XXIV

From a mixture of the coloured products formed, we succeeded in the identification[54] of dimer XXV and trimer XXVI and in the isolation of an analogous tetramer. The connection existing between the colourless oligomers XXI and XXII with the coloured oligomers XXV and XXVI has been proved. The compounds causing discolouration of the polymer lose their colour in a reaction with an excess of alkyl hydroperoxide; however, no products of type XXIII are formed in this case.

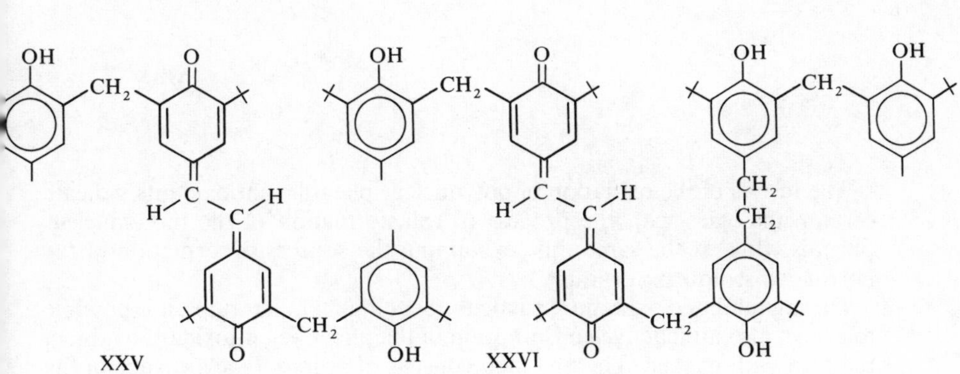

XXV

XXVI

219

It has been found[55], by gel chromatographic investigation of the reaction mixture obtained by acting with tert-butyl hydroperoxide on 2,2'-methylene-bisphenols analogous to 2,2'-methylenebis(4-methyl-6-tert-butylphenol) that oligomeric products similar to XXI and XXII are probably also formed from a derivative having an ethyl group at the position para to the hydroxyl, while the derivative substituted with the tert-butyl group reacts quite differently (the formation of a compound of type XVI should rather be expected).

Cyclohexadienone XXVII is formed by oxidation of another technically important antioxidant, 1,3,5-trimethyl-2,4,6-tris-(3,5-di-tert-butyl-4-hydroxy-benzyl)benzene with tert-butylperoxyl[50]. The thermal stability of this compound is limited. At a temperature above 160°C, brown products and volatile compounds are formed therefrom; of these, acetone and 2,6-di-tert-butyl-1,4-benzoquinone could be identified, and the presence of the compounds of type XXVIII and XXIX is supposed on the basis of the mass spectra.

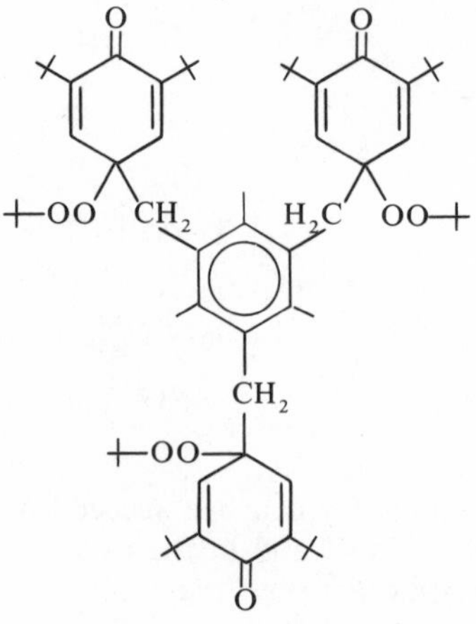

XXVII

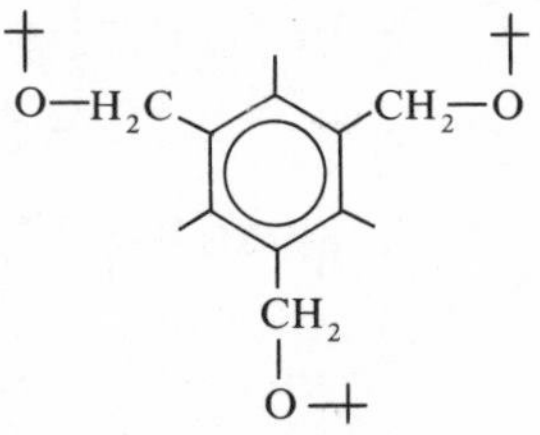

XXVIII

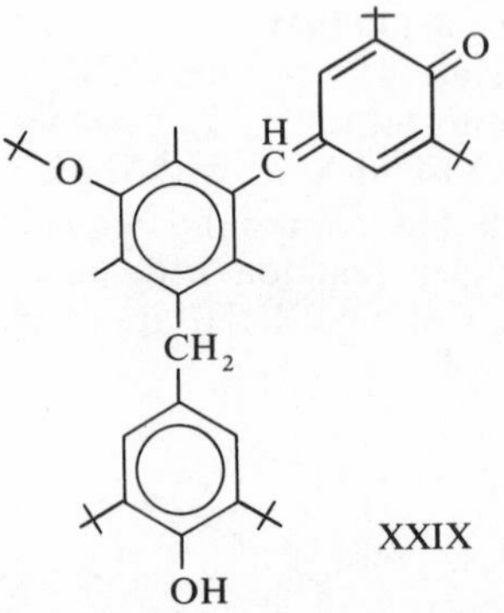

XXIX

The results of the oxidation of polynuclear phenolic antioxidants indicate certain analogies with the process of transformation of the mononuclear phenols while at the same time explaining the sources of formation of the polymer-discolouring products.

During inhibited oxidation, particularly at high temperatures, it is possible to expect also an inactive consumption of the phenolic antioxidant by direct reaction with oxygen. The first step consists of a homolytic cleavage of the

H—O bond and of the formation of an aryloxyl. It is more likely, however, that this aryloxyl formation is due to a reaction with the ROO˙ radical, followed by a reaction with oxygen at a rate depending on the stability of aryloxyl[10, 11, 56]. For aryloxyls derived from the majority of mononuclear phenolic antioxidants the formation of cyclohexadienoneperoxyl (XXX) followed by the formation of cyclohexadienone peroxide (XXXI) can be assumed.

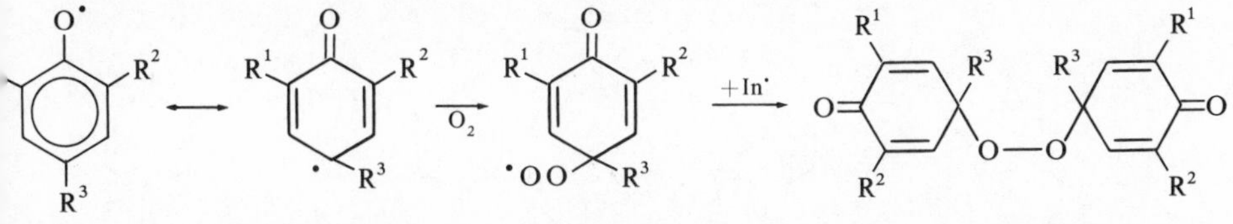

XXX   XXXI

The peroxides of type XXXI are thermally and photochemically relatively unstable, slightly coloured compounds. It is therefore not possible to expect that they may be identified during the transformation of the antioxidants in the process of ageing of the polymers and particularly during thermo-oxidative degradation. However, their transient formation and the homolysis of the O—O bond can—similarly to the case of alkylperoxycyclohexadien-ones—explain the formation of some products of the oxidative transforma-tion. An example can be seen in the isolation of 2,6-di-tert-butyl-1,4-benzo-quinone as a product of the oxidation of 2,6-di-tert-butyl-4-alkoxyphenol[57,58]. A mixture of compounds has been formed[50] by the direct oxidation of 1,3,5-trimethyl-2,4,6-tris-(3,5-di-tert-butyl-4-hydroxyphenol)benzene with oxygen at 230°. In this mixture, besides mononuclear products probably due to the thermal cleavage of the cyclohexadienone derivative XXVII (i.e. 2,6-di-tert-butyl-1,4-benzoquinone and 3,5-di-tert-butyl-4-hydroxybenzalde-hyde), quinone methide XXXIII was identified by the mass spectrum.

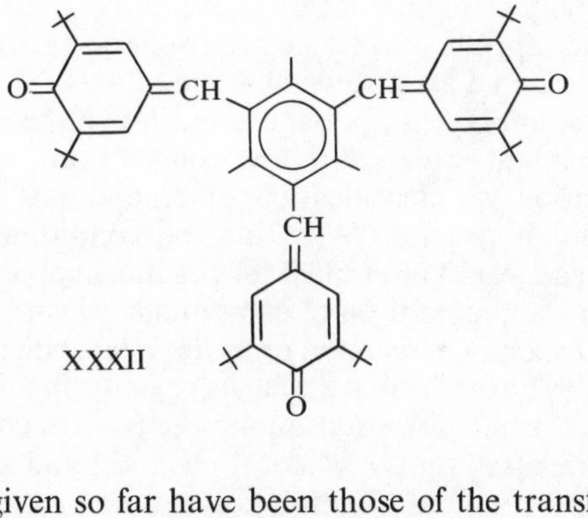

XXXII

The examples given so far have been those of the transformations of the antioxidants from the group of the monohydric mono- or polynuclear phenols substituted with alkyls or alkoxyls. Dihydric phenols represent another

221

group of active antioxidants. They are practically exclusively mono-tert.-alkylated derivatives of pyrocatechol and hydroquinone. Their transformation yields, in the first place, benzoquinoid (XXXIV, XXXV) and hydroxybenzoquinoid (XXXVI, XXXVII) compounds. The synthesis of both isomeric benzoquinones is easy[59]. Tertiary alkylated hydroxybenzoquinones were obtained by the oxidation of mono-tert-alkyl-pyrocatechols and hydroquinones with oxygen in an alkaline medium[60,61].

XXXIII   XXXIV   XXXV   XXXVI

By means of e.p.r., useful data on the mechanism of the autoxidation process were obtained[62]. Benzoquinones and hydroxybenzoquinones were also obtained by the reaction of 4-tert-butylpyrocatechol with tert-butylperoxyl, tert-butyl-hydroperoxide, and tetralylhydroperoxide; some further unidentified compounds were formed in this process.

From 2-tert-butylhydroquinone, only corresponding 2-tert-butyl-1,4-benzoquinone was formed[63]. Owing to the comparatively high thermal instability of benzoquinones, no simple quinoid products except for 2-tert-butyl-1,4-benzoquinone could be identified unambiguously between products formed by the oxidation of mono-tert-butylated pyrocatechol and hydroquinone with oxygen at 150°C in trichlorobenzene solution[64]. According to TLC, some other products due to both coupling and cleavage reactions are present in the mixture along with hydroxybenzoquinones. It has been deduced from the investigation of the transformations of dihydric phenolic antioxidants that the deactivation by oxidation with oxygen occurs at a measurable rate only at temperatures above 100°C. Therefore, such transformation does not appear during the atmospheric ageing of the polymers, but it must be considered at processing temperatures.

For the stabilization of both polymers and low-molecular weight hydrocarbons the antioxidants are used in low concentrations. This renders very difficult the isolation or identification of compounds formed therefrom during the inhibition process. As a rule the conditions of the oxidative degradation, such as would be studied for practical applications, are not the optimum ones for the preparation of compounds which have been obtained in a defined form by a model reaction or by an independent synthesis. Consequently, it must be borne in mind that in practice incomparable data are already obtained at small deviations under the process conditions.

Most results are based on the classical chemical and spectral analysis of laboriously isolated products. The application of chromatographic separation and identification techniques makes the task somewhat easier, but is by no means an absolute solution, as long as it cannot guarantee an uncompromisingly careful treatment of the concentrates of the transformation products, and if— and this is one of the main difficulties of the whole problem—there is not at our disposal a sufficient selection of standard compounds modelling the products of the primary transformation, as well as compounds obtained therefrom as secondary ones by further oxidation and by a thermal or photochemical process. An investigation of the transformations by means of standards is extremely labour-consuming, but it is the only one real approach to the solution of the problem.

So far, not many data have been reported about the transformation products of antioxidants isolated and identified during inhibited oxidation. Owing to high oxidation temperatures, in the first place more detailed data about the products formed during the oxidative degradation of polymers are missing. However, compounds detected in stabilized liquid hydrocarbon substrates oxidized in a temperature range to 100° are indicative for the process of the atmospheric ageing of polymers. Along with the knowledge of physical properties and reactivity of the identified compounds, they allow certain extrapolations for high-temperature degradation. A difficulty is caused by the formation of undefined 'tarlike' compounds, into which an essential part of the antioxidant is often transformed. Some data reported on the transformation products must be accepted with a certain reserve, since some of the compounds isolated may also be those formed only during the process of isolation.

Data were published dealing with the transformations of methylated phenols. Para-cresol and 2,4- and 2,6-xylenols present in autoxidized cumene changed[65] to products of the C—C dimerization. Diphenoquinone (III, R^1, R^2 = Me) was formed during the autoxidation of benzaldehyde stabilized with 2,6-xylenol[66].

The majority of papers published so far have been devoted to 2,6-di-tert-butyl-4-methylphenol, an antioxidant that is widely used in practice. The investigation is facilitated by good knowledge of the structure of the products formed by model reactions and by the generally unambiguous reactivity of this phenol. It was changed in oxidized cumene[27, 43, 67] into the dihydroxy-dibenzyl derivative (V) and stilbenequinone (VI). During the stabilization of a lubricating oil, the formation of an aldehyde, probably 3,5-di-tert-butyl-4-hydroxybenzaldehyde, was observed[68]. By means of a chromatographic analysis of the transformation products formed at the end of the induction period during the autoxidation of tetralin initiated with azobisisobutyronitrile, only the presence of peroxycyclohexadienones[47] (XIX, R^1, R^2 = t-Bu, R^3 = Me), obtained by a reaction with the ROO˙ radicals derived from the oxidized substrate and from the initiator was proved. No products having the stilbenequinoid structure have been observed in this case. The cyclohexadienones isolated in the latter case confirm the kinetic suggestions of the course of this inhibition reaction[69]. However, the data on the transformations of 2,6-di-tert-butyl-4-methylphenol in various substrates indicate that it is very difficult to outline a generally valid scheme of transformation also in the case of a very simple antioxidant used under relatively mild conditions.

There are not many chemical data about the transformations of phenols in polymers. It has been found[70] that during the oxidation of isotactic polypropylene stabilized with 2,4,6-tri-tert-butylphenol, cyclohexadienone peroxide was obtained by a sequence of reactions which are assumed to proceed via the formation of aryloxyl, its interaction with polypropylene hydroperoxide, and further transformations. 4,4'-Methylene-bis-(2,6-di-tert-butylphenol) is believed to react in a similar way.

From 2,6-di-tert-butyl-4-methylphenol, used for the stabilization of isotactic polypropylene, some products were formed during the oxidation carried out at 200°, of which 3,5-di-tert-butyl-4-hydroxybenzaldehyde and 2,2',6,6'-tetra-tert-butyl-4,4'-stilbene-quinone[71] were identified.

The transformations of another important antioxidant, 2,2'-methylene-bis-(4-methyl-6-tert-butylphenol), during the polyolefin stabilization are assumed[72–76] on the basis of kinetic data. However, the chemical nature of the products was not proposed. According to ref. 77 a multicomponent system is formed. We used the TLC and GPC methods to investigate[54, 78, 79] the transformations of 2,2'-methylene-bis-(4-methyl-6-tert-butylphenol) during the oxidation of tetralin and polypropylene. Owing to the lower temperature of the tetralin oxidation (60°C), it was possible to prove the formation of a tetralyl analogue of alkylperoxycyclohexadienone XXIII in addition to the phenolic products of oxidative coupling (dimer XXI and trimer XXII) and of a mixture of brown-coloured products in which trimer XXVI was the main component. In the case of the oxidation of atactic (150°C) and isotactic (180°C) polypropylene, brown products are formed from the onset of the oxidation. However, their exact identification causes difficulties, especially in isotactic polypropylene, since owing to the confirmed thermal instability of oligomers with the stilbenequinoid structure they readily undergo further changes. Dimers XXI and XXV are proved[79] in atactic polypropylene by means of TLC. The presence of low amounts of phenolic oligomers XXI and XXII cannot be excluded in the oxidized isotactic polypropylene; these oligomers are rapidly consumed by an inhibition reaction[78] at 180°C, which is the reason why their momentary concentration is so low. If also the peroxidic derivative of cyclohexadienone is transiently formed at high oxidation temperatures, it undergoes thermal splitting and can be the source of compounds having a lower molecular weight than the original methylene bisphenol. Such compounds have been observed[80] by means of GPC in stabilized tetralin oxidized at temperatures above 100°C. Such compounds can also arise from products having the stilbenequinoid structure in the oxidation steps following the induction period, i.e. at a higher content of alkyl peroxyls or alkyl hydroperoxides and after consumption of the radical scavengers (either the methylenebisphenol originally present in the mixture or the products arising therefrom and exhibiting antioxidative activity). The benzoquinone derivatives may be present among these compounds, as has been demonstrated[81] by a model investigation with tetra-tert-butylstilbenequinone.

Interesting data on the structure of the products obtained by a transformation of the antioxidant during the oxidation of polypropylene stabilized with 1,3,5-tri-methyl-2,4,6-tris-(3,5-di-tert-butyl-4-hydroxybenzyl)benzene have been reported[71]. Owing to the fact that each of the hydroxybenzyl groups bound in this antioxidant to the central benzene ring can react independently, a very complex mixture of the oxidation products is formed; thirteen compounds have been identified by mass spectroscopy. They include

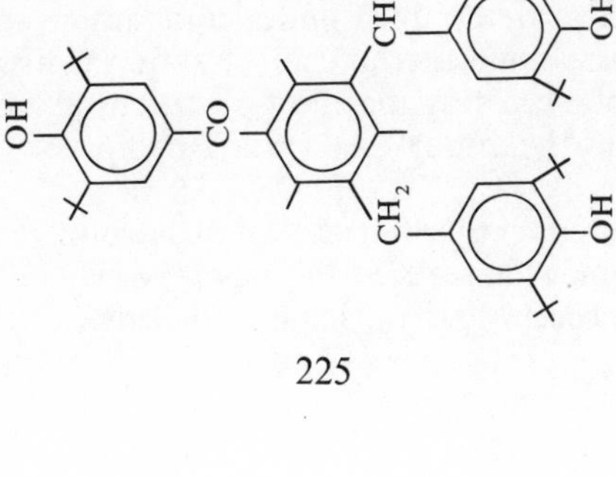

XL

XXXIX

XXXVIII

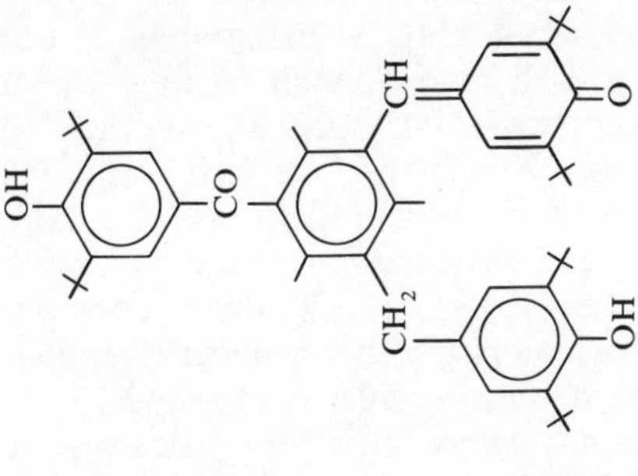

XXXVII

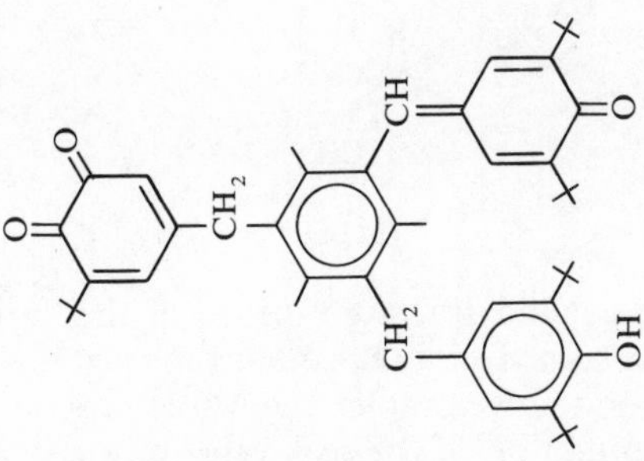

225

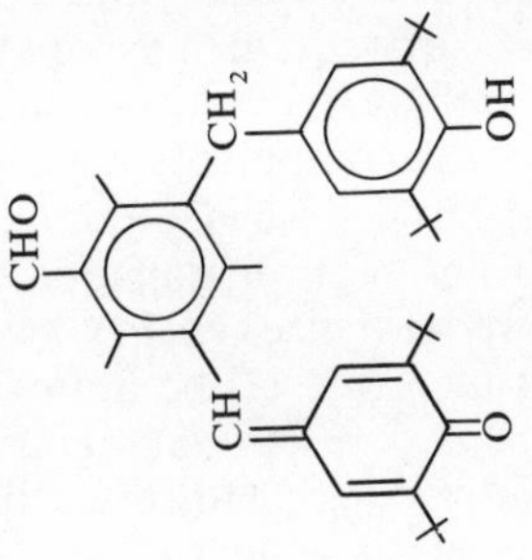

compounds oxidized on the individual rings to a different degree. Some of the products contain the carbonyl groups, quinone methide or 1,2-benzoquinoid structures, as well as structures obtained by the oxidative splitting-off of the benzyl group, e.g. compounds XXXVII to XL.

However, 2,6-di-tert-butyl-1,4-benzoquinone was the main product of this transformation. It has been demonstrated[82], by using 1,3,5-trimethyl-2,4,6-tris-(3,5-di-tert-butyl-4-hydroxybenzyl)benzene labelled with ^{14}C, that a part of the oxidized molecule of the antioxidant is bonded to the polypropylene molecules. Structure XLI analogous to product XXVIII from the model reaction is considered here. However, the existence of compound XLI seems somewhat uncertain, owing to the steric requirements of such a molecule.

$$PP-O-H_2C \underset{CH_2-O-PP}{\overset{CH_2-O-PP}{\bigcirc}} CH_2-O-PP$$

XLI

The data on the simulated transformations of the antioxidants indicate along with the data on some isolated products formed under real conditions of the inhibited oxidation how complex the whole process is. They point out the difficulty of determination of the real value of the stoichiometric factor giving the number of peroxyls deactivated by one molecule of a phenolic antioxidant based on the chemical analysis of the products. A more probable value of the stoichiometric factor determined on the basis of the products obtained can be approached only by using some antioxidants which are particularly suited for such study owing to their unequivocal reactivity. 2,6-Di-tert-butyl-4-methylphenol is such a compound. In the case of an investigation of the antioxidants which react in a more complex way while yielding compounds some of which undergo further oxidation immediately after their formation, the determination of the stoichiometric factor based on the products is rather illusory. This group of compounds includes polyhydric phenols and the majority of polynuclear phenolic antioxidants used for the stabilization of macromolecular compounds.

Also the value of the stoichiometric factor determined from the kinetic data should be regarded as informative only, since it is an integral fact involving the results of a whole complex of reactions occurring in varied ratio during the process of inhibition. These processes play their role in the kinetic evaluation as a complete set which is difficult to separate. However, the stoichiometric factor determined under comparable conditions for antioxidants for which the same mechanism of action can be considered is a useful qualitative tool indicating the effect of structural factors or the extent of the participation of the antioxidants being compared in the undesirable side reactions[84].

It can be said, consequently, that the knowledge of the transformations of the antioxidants somewhat restricts the general validity of a kinetic analysis of the stabilization process and puts the knowledge of the mechanism of action of the antioxidants on a more realistic basis indicating more accurately the situation arising during atmospheric ageing and the degradation of polymers. At the same time, it allows information to be obtained about the participation of the transformation products in the individual processes occurring during ageing. A mere survey of the structures of compounds which can be formed under certain conditions from the originally present antioxidants in the stabilized polymer indicates that the compounds involved cannot always be inert in the further process of degradation. Their effect on the further course of oxidation is widely affected by the conditions of the process. I shall outline here the knowledge acquired so far about the influence of the individual types of transformation products on the process of autoxidation.

During the oxidation inhibited with monohydric mononuclear phenols[47] or polynuclear phenols[50, 53], alkylperoxycyclohexadienones are formed, if the temperature of oxidation does not exceed the temperature of their thermal decomposition. We investigated[85] the effect of the simplest compounds of this series having the structure

where R = tert-butyl, cumyl, α-tetralyl, or 2-cyanoisopropyl, in the course of the oxidation of tetralin. At 65°C, the presence of these alkylperoxycyclohexadienones has no influence on the rate of oxidation. However, an analogous alkylperoxycyclohexadienone XXIII derived from 2,2′-methylenebis-(4-methyl-6-tert-butylphenol) had a weak retardative effect[78]. The homolysis of the O—O bond at an elevated temperature raises the effect of these compounds, and at 150° (i.e. above the point of incipient decomposition of all the compounds under investigation) there is a clear pro-oxidative effect. The ability of the RO˙ radicals produced by homolysis to initiate or propagate the autoxidative chains will play its role here. This process must be taken into account during processing or thermal treatment of polymers in which some alkylperoxycyclohexadienones were present as a result of transformations during preceding low-temperature oxidation. The products of a further thermal or oxidative transformation of these cyclohexadienones, 3,5-di-tert-butyl-4-hydroxybenzaldehyde or 2,6-di-tert-butyl-1,4-benzoquinone, have no antioxidative properties[82, 86, 87].

Benzoquinones and hydroxybenzoquinones have been proved to be products of transformations of pyrocatechol, and benzoquinones products of hydroquinone antioxidants[63]. They are effective inhibitors of polymerization[88]. They scavenge the R˙ radicals, and their presence can—in the first place—be reflected in the onset of oxidation, since their interference with the propagation step 2 can be assumed here, above all, at an insufficient pressure of oxygen. Kinetic treatment of the oxidation process indicates the possibility of such a retardative effect[66, 89–91]. We studied the behaviour of a series of isomeric mono-tert-alkylated benzoquinones and hydroxybenzoquinones XXXIII to XXXVI during initiated oxidation of tetralin[87] and found that with the exception of 2-tert-butyl-1,4-benzoquinone, the rate of the oxidation remained unaffected by any of the alkyl-1,4-benzoquinones and hydroxy-1,4-benzoquinones investigated. The same results were obtained when investigating the properties of (2-alkoxyalkyl)-1,4-benzoquinones[92]. However, 4-tert-alkyl-1,2-benzoquinones exhibited retardative activity. The different behaviour of isomeric benzoquinones and the semiquinones preceding them can also be one of the causes of the different antioxidative activity of the derivatives of the pyrocatechol and hydroquinone series.

Some changes in the action of the quinoid products can take place in the process of the atmospheric ageing of polymers as a consequence of the radiation co-effect. From inactive tert-alkyl-1,4-benzoquinones, (2-alkoxyalkyl)hydroquinones can be formed[93] in the presence of alcohols (which are found among the secondary oxidation products of hydrocarbons); for these (2-alkoxyalkyl)hydroquinones, a higher antioxidative activity has been proved during the stabilization of isotactic polypropylene and tetralin than that of tert-alkylated hydroquinones from which they are derived[92]. The transformation of 2-tert-butylhydroquinone is given as an example. 2,5- and 2,6-Di-tert-alkylhydroquinones also undergo gradual changes in a similar way.

The oxidative transformations of 2,2'-methylene-bis-(4-methyl-6-tert-butylphenol) give rise to compounds with antioxidative activity. Their ability to stabilize to a certain degree various macromolecular substrates is indicated[76, 77, 94, 95] without any knowledge of their structure. However, the transformation of this antioxidant is also considered to be the source of free radicals, acting as initiators during the oxidation of polypropylene[96]. We have investigated in detail well-defined model compounds during the stabilization of isotactic polypropylene and tetralin and have demonstrated the antioxidative activity of the main individual transformation products and of the synthetically prepared mixtures thereof[78]. At 60°C (tetralin oxidation), the antioxidative properties of the original methylene-bis-phenol and its dimer XXI did not differ, while trimer XXII was somewhat less active. Surprisingly, the mixture of the original methylene-bis-phenol and cyclo-hexadienone XXIII had a synergistic effect.

Good antioxidative properties—although lower than those of the original methylene-bis-phenol—have been found for phenolic products of the oxida-tive coupling (XXI and XXII) in the stabilization of isotactic polypropylene. The antioxidative activity gradually decreases when being recalculated to the same relative content of the hydroxyl groups from the monomeric phenol to the trimer. However, no synergism between the phenolic transformation products is observed in this case, similarly to tetralin. Owing to the presence of phenolic hydroxyls in the coloured transformation products, the latter also preserve a certain antioxidative activity. Alkylperoxycyclohexadienone XXIII added to polypropylene did not even reduce its induction period, although the oxidation proceeded above its decomposition temperature. However, this cyclohexadienone derivative has a distinctly negative effect in a mixture with the original 2,2'-methylene-bis-phenol.

I would also like to mention here the retardative activity of the transforma-tion products of phenolic antioxidants substituted with sulphur-containing groups and thiobis-phenols which appears in the step following the induction period[86, 97]. The retardative activity observed during the stabilization of polypropylene is explicitly a consequence of the presence of the sulphur atoms in the molecule. The character of the transformation products has not been exactly defined.

A complex knowledge of the properties of the transformation products of phenolic antioxidants allows us to evaluate more realistically the mechanism of the antioxidative effect under the conditions of polymer ageing. A large number of authors have contributed independently to the characterization of the products. Isolated products show that after the primary reaction between the antioxidant and the alkylperoxyl 4, further processes follow which are dependent on the conditions of degradation. These processes need not correspond to the course taking place under simulated model conditions; therefore, stoichiometry of the mutual interactions between the antioxidant and the species propagating the kinetic chain is not nearly so simple as could be assumed from pure kinetic studies. In the case of a practical application to the mechanism of action of the antioxidants during the ageing and thermal degradation of polymers, a whole complex of effects play their role in a ratio that is difficult to define, involving—apart from comparatively well-known oxidative and thermal transformations of the antioxidants—also the co-effect of the ions of metals present as impurities or as the residues of the poly-merization catalysts, radiation-induced photochemical reactions, as well as mechanical effects that cannot be neglected either. The products that are formed from the components of the stabilizing mixture in the complicated scheme of the parallel and consecutive reactions are accumulated in the stabilized polymer and affect its stability and appearance depending on their character. Therefore, a mere exact knowledge of the relationships between the structure and effectivity of the antioxidants is insufficient for a complex characterization of the applicability of the antioxidants to specific conditions, and the knowledge of the chemistry of transformations must necessarily become more effective. These transformations are valid not only for the region of the investigation of polymer stabilization, but also, with small modifications only, for the stabilization of further technically important organic substrates, such as gasolines, oils, and lipidic substrates in the first place. They also have a potential importance for the study of the interactions of rather complicated systems having the properties of the radical scavengers in the biological systems *in vivo*, especially in the investigation of the processes of biological ageing of humans. It is just for this region, where the synthetic polymers and biopolymers contact each other, that the investigation of the chemical transformations of antioxidants must be stimulated.

REFERENCES

[1] W. O. Lundberg (Editor), *Autoxidation and Antioxidants*, Volumes I and II. Interscience: New York and London (1961 and 1962).
[2] J. Voigt, *Die Stabilisierung der Kunststoffe gegen Licht und Wärme*. Springer: Berlin (1966).
[3] G. Scott, *Atmospheric Oxidation and Antioxidants*. Elsevier: Amsterdam, London and New York (1965).
[4] J. Pospíšil, *Antioxidanty*. Academia: Praha (1968).
[5] M. B. Neiman (Editor), *Aging and Stabilization of Polymers*. Consultants Bureau: New York (1965).
[6] K. Thinius, *Stabilisierung und Alterung von Plastwerkstoffen*. Akademie-Verlag: Berlin (1969).
[7] W. L. Hawkins (Editor), *Polymer Stabilization*. Wiley-Interscience: New York and London (1972).
[8] J. Pospíšil, *Kinetics and Mechanism of Polyreactions*, IUPAC International Symposium on Macromolecular Chemistry, Plenary and Main Lectures Volume, p 789, Budapest (1969).
[9] W. R. Hatchard, R. D. Lipscomb and F. W. Stacey, *J. Amer. Chem. Soc.* **80**, 3636 (1958).
[10] E. R. Altwicker, *Chem. Rev.* **67**, 475 (1967).
[11] A. R. Forrester, J. M. Hay and R. H. Thomson, *Organic Chemistry of Stable Free Radicals*. Academic Press: London and New York (1968).
[12] D. Braun, 'Polyradicals', in H. F. Mark, N. G. Gaylord and N. M. Bihales (Editors), *Encyclopedia of Polymer Science and Technology*, Vol. 15, p 429. Interscience: New York (1971).
[13] E. Müller, K. Ley and W. Kiedaisch, *Chem. Ber.* **87**, 1605 (1954).
[14] A. J. Waring, 'Cyclohexadienones', in H. Hart and G. J. Karabatsos (Editors), *Advances in Alicyclic Chemistry*, Vol. I, p 129. Academic Press: New York (1966).
[15] J. Petránek and J. Pilař, *Coll. Czech. Chem. Commun.* **34**, 79 (1969).
[16] J. Petránek and J. Pilař, *Coll. Czech. Chem. Commun.* **35**, 830 (1970).
[17] D. G. Hewitt, *J. Chem. Soc. C*, 2967 (1971).
[18] E. Müller, H. Kaufmann and A. Rieker, *Liebig's Ann.* **671**, 61 (1964).
[19] G. D. Cooper, H. S. Blanchard, G. F. Endress and H. Finkbeiner, *J. Amer. Chem. Soc.* **87**, 3996 (1965).
[20] C. C. Price and N. Nakaoka, *Macromolecules*, **4**, 363 (1971).
[21] A. Rieker, N. Zeller, K. Schurr and E. Müller, *Liebig's Ann.* **697**, 1 (1966).
[22] R. H. Bauer and G. M. Coppinger, *Tetrahedron*, **19**, 1201 (1963).
[23] R. Magnusson, *Acta Chem. Scand.* **18**, 759 (1964).
[24] C. D. Cook and B. E. Norcross, *J. Amer. Chem. Soc.* **78**, 3797 (1956).
[25] C. Cook and N. D. Gilmour, *J. Org. Chem.* **25**, 1429 (1960).
[26] R. Stubbins and F. Sicilio, *Tetrahedron*, **26**, 291 (1970).
[27] D. H. Hey and W. A. Waters, *J. Chem. Soc.* 2754 (1955).
[28] A. L. Buchachenko, *Stabilnyje Radikaly*, Izd. Akad. Nauk SSSR: Moskva (1963).
[29] G. M. Coppinger, *J. Amer. Chem. Soc.* **86**, 4385 (1964).
[30] E. A. Chandross, *J. Amer. Chem. Soc.* **86**, 1263 (1964).
[31] N. C. Yang and A. J. Castro, *J. Amer. Chem. Soc.* **82**, 6208 (1960).

[32] E. A. Chandross and R. Kreilick, *J. Amer. Chem. Soc.* **85**, 2530 (1963); **86**, 117 (1964).
[33] Y. Yoshikawa and J. Kumanotani, *Makromol. Chem.* **134**, 33 (1970).
[34] J. Bourdon and M. Calvin, *J. Org. Chem.* **22**, 101 (1957).
[35] E. Müller, R. Mayer, B. Narr, A. Rieker and K. Scheffler, *Liebig's Ann.* **645**, 25 (1961).
[36] E. Müller, A. Schick, R. Mayer and K. Scheffler, *Chem. Ber.* **93**, 2649 (1960).
[37] E. Müller, H. B. Stegmann and K. Scheffler, *Liebig's Ann.* **645**, 79 (1961); **657**, 5 (1962).
[38] J. Petránek, *Tetrahedron*, **27**, 5201 (1971).
[39] H. D. Becker, *J. Org. Chem.* **29**, 3068 (1964).
[40] E. Müller, A. Rieker and A. Schick, *Liebig's Ann.* **673**, 40 (1964).
[41] T. Matsura and A. Nishinaga, *J. Org. Chem.* **27**, 3072 (1962).
[42] L. R. Mahoney and A. A. DeRooge, *J. Amer. Chem. Soc.* **89**, 5619 (1967).
[43] A. F. Bickel and E. C. Kooyman, *J. Chem. Soc.* 3211 (1953).
[44] T. W. Campbell and G. M. Coppinger, *J. Amer. Chem. Soc.* **74**, 1469 (1952).
[45] W. A. Pryor, D. M. Huston, T. R. Fiske, T. L. Pickering and E. Cinfarrin, *J. Amer. Chem. Soc.* **86**, 4237 (1964).
[46] R. A. Sheldon and J. K. Kochi, *J. Org. Chem.* **35**, 1223 (1970).
[47] E. C. Horswill and K. U. Ingold, *Canad. J. Chem.* **44**, 263 (1966).
[48] E. C. Horswill and K. U. Ingold, *Canad. J. Chem.* **44**, 269 (1966).
[49] H. Berger and A. F. Bickel, *Trans. Faraday Soc.* **57**, 1325 (1961).
[50] J. Koch, *Angew. Makromol. Chem.* **20**, 7 (1971).
[51] H. R. Gersmann and A. F. Bickel, *Proc. Chem. Soc.* 231 (1957).
[52] M. S. Kharash and B. S. Joshi, *J. Org. Chem.* **22**, 1435 (1957).
[53] L. Taimr, H. Pivcová and J. Pospíšil, *Coll. Czech. Chem. Commun.* **37**, 1912 (1972).
[54] L. Taimr and J. Pospíšil, IUPAC International Symposium on Macromolecules, Helsinki, July 1972; *Angew. Makromol. Chem.*, **28**, 13 (1973).
[55] L. Zikmund, J. Brodilová and J. Pospíšil, 11th IUPAC Microsymposium on Macromolecules, Prague, September 1972.
[56] A. G. Davies, *Organic Peroxides*. Butterworths: London (1961).
[57] E. Müller and K. Ley, *Chem. Ber.* **88**, 601 (1955).
[58] C. D. Cook and R. C. Woodworth, *J. Amer. Chem. Soc.* **75**, 6242 (1953).
[59] J. Cason, 'Synthesis of benzoquinones by oxidation', in R. Adams and others (Editors), *Organic Reactions*, Vol. IV, p 305. Wiley: New York (1948).
[60] J. Pospíšil and V. Ettel, *Coll. Czech. Chem. Commun.* **24**, 341 (1959).
[61] I. Buben and J. Pospíšil, *Coll. Czech. Chem. Commun.* **34**, 1991 (1969).
[62] J. Pilař, I. Buben and J. Pospíšil, *Coll. Czech. Chem. Commun.* **35**, 489 (1970).
[63] J. Pospíšil, J. Horák and L. Taimr, IUPAC International Symposium on Macromolecules, Helsinki, July 1972.
[64] J. Lerchová, M. Obali and J. Pospíšil, 11th IUPAC Microsymposium on Macromolecules, Prague, September 1972.
[65] R. F. Moore and W. A. Water, *J. Chem. Soc.* 243 (1954).
[66] R. F. Moore and W. A. Water, *J. Chem. Soc.* 2432 (1952).
[67] A. F. Bickel and E. C. Kooyman, *J. Chem. Soc.* 2415 (1957).
[68] J. I. Wasson and W. M. Smith, *Industr. Engng Chem.* **45**, 197 (1953).
[69] C. Boozer, G. S. Hammond, C. E. Hamilton and J. W. Sen, *J. Amer. Chem. Soc.* **27**, 3233 (1955).
[70] E. G. Sklyarova, A. F. Lukovnikov, M. L. Khidekel and V. U. Karpov, *Izvest. Akad. Nauk SSSR*, Ser. Khim. 1093 (1965).
[71] J. Koch, *Angew. Makromol. Chem.* **20**, 21 (1971).
[72] Ju. A. Shlyapnikov, V. B. Miller and E. S. Torsueva, *Izvest. Akad. Nauk SSSR*, 1966 (1961).
[73] Ju. A. Shlyapnikov, V. B. Miller, M. B. Neiman and E. S. Torsueva, *Vysokomol. Soedin.* **5**, 1507 (1963).
[74] Ju. A. Shlyapnikov, V. B. Miller, M. B. Neiman and E. S. Torsueva, *Dokl. Akad. Nauk SSSR*, **151**, 148 (1963).
[75] Ju. A. Shlyapnikov and V. B. Miller, *Zh. Fiz. Khim.* **39**, 2814 (1965).
[76] I. Gömöry, J. Štímel, R. Reya and A. Gömöryová, *J. Polym. Sci. C*, **16**, 451 (1967).
[77] I. Gömöry, A. Gömöryová, R. Reya and J. Štímel, *Europ. Polym. J.*, Suppl. 545 (1969).
[78] L. Zikmund, L. Taimr, J. Čoupek and J. Pospíšil, *Europ. Polym. J.* **8**, 83 (1972).
[79] M. Prusíková and J. Pospíšil, unpublished results.
[80] L. Zikmund, unpublished results.
[81] L. Taimr, unpublished results.
[82] J. Koch and K. Figge, *Angew. Makromol. Chem.* **20**, 35 (1971).

[83] M. Prusíková, L. Jiračková and J. Pospíšil, *Coll. Czech. Chem. Commun.*, **37**, 3788 (1972).

[84] E. Lisá and J. Pospíšil, 11th IUPAC Microsymposium on Macromolecules, Prague, September 1972.

[85] I. Buben and J. Pospíšil, unpublished results.

[86] L. Jiráčková and J. Pospíšil, IUPAC International Symposium on Macromolecules, Helsinki, July 1972; *Europ. Polym. J.*, in press.

[87] J. Pospíšil, E. Lisá and I. Buben, *Europ. Polym. J.* **6**, 1347 (1970).

[88] H. Hartel, *Chimia*, **19**, 116 (1965).

[89] J. R. Dunn, W. A. Waters and C. Wickam-Jones, *J. Chem. Soc.* 242 (1952).

[90] J. W. Breitenbach, O. F. Olaj and A. Schindler, *Monatsh. Chem.* **88**, 1115 (1958).

[91] V. V. Guryanova, B. M. Kovarskaya, V. B. Miller and J. A. Shlyapnikov, *Izvest. Akad. Nauk SSSR*, Ser. Khim. 289 (1971).

[92] E. Lisá, L. Kotulák, J. Petránek and J. Pospíšil, *Europ. Polym. J.* **8**, 501 (1972).

[93] J. Petránek, O. Ryba and D. Doskočilová, *Coll. Czech. Chem. Commun.* **32**, 2140 (1967).

[94] I. Gömöry, *Europ. Polym. J.* **6**, 1047 (1970).

[95] I. Gömöry and A. Gömöryová, IUPAC International Symposium on Macromolecules, Helsinki, July 1972.

[96] A. A. Berlin, A. A. Ivanov and A. P. Firsov, *Vysokomol. Soedin.* **13A**, 2713 (1971).

[97] L. Jiráčková and J. Pospíšil, *Europ. Polym. J.* **8**, 75 (1972).

THE DEVELOPMENT OF METHODS FOR EXAMINING STABILIZERS IN POLYMERS AND THEIR CONVERSION PRODUCTS

ELISABETH SCHROEDER

Technische Hochschule für Chemie 'Carl Schorlemmer', Department of High Polymers, Leuna-Merseburg, GDR

ABSTRACT

In the field of stabilizer analysis today there is a need for the analytical determination of the decomposition or transformation products of the stabilizers in polymer systems including their reaction products with the polymers. For the solution of such a complex analytical task, separation problems are at the centre of modern scientific work. These separation problems are subdivided into two partial processes, namely (a) separation of the low molecular substances from the polymer; (b) separation of the mixture of low molecular substances.

Selective stage extractions and gel permeation chromatography are particularly favourable. The well-known method of the precipitation of the polymer from diluted solutions needs a more thorough scientific consideration of the solvent–precipitant effects on the stability of the reaction products of stabilizers or their fragments with the polymer.

Thin-layer chromatography (TLC) has brought the greatest success, multistage processes here also offer prospective possibilities for the separation of complex systems. Gas chromatography (GC) is less important. Liquid chromatography (LC) with new high-efficiency carrier materials having particle diameters in the micro range will be most important in the future.

Structure examinations of the decomposition products of the stabilizers in principle correspond to the classical organo-analytical concepts after separation processes. The most important of the decomposition products of the stabilizers were those of the tin compounds and those of stabilizers based on urea. All toxic substances were detectable by TLC.

Structure examinations of stable transformation products of the nitrogen-containing antioxidants with u.v. and i.r. spectrometry are demonstrated in relation to the most discussed mechanism of the inhibition reaction. An analytical scheme for oxidized vulcanizates in the presence of secondary and tertiary mono- and diamines is given at the end to demonstrate the necessity of combining largely selective separation processes with all analytical detection and determination processes of high sensitivity for a successful overall analytical process.

1. INTRODUCTION

On the one hand, progress in the field of stabilizer analytical techniques is closely related to progress in all analytical disciplines—especially in the

233

instrumental techniques—and on the other it is a reflection of the growing standard of knowledge in the field of the degradation of polymers and their inhibition. The trend of analytical chemistry is less characterized by the search for new measuring principles, but it indicates a preferred accentuation of the increase in the measuring sensitivity, selectivity, reproducibility and speed, of which the most important at present are the increase in sensitivity and selectivity for the stabilizer techniques. While in the 'fifties research work here was concentrated on the evidence and quantitative determination of stabilizers or their mixtures in polymers, today there is the demand for analytical determination of the decomposition or conversion products of the stabilizers in polymer systems including their reaction products with the polymers. This task, which is for the analyst both difficult and at the moment only conditionally soluble, results from the high demands made by all of us everywhere on the toxic purity of a plastic material used as a commodity, as well as from the necessity of an analytical backing for explaining the mechanism of action of stabilizers when they are used in polymers to prevent their destruction by heat, oxidation, light or by the combined effect of these and other factors. The field of work for stabilizer techniques thus arises not only from scientific but also from economic points of view and confronts scientists who seek to protect human health against noxious environmental influences.

2. ACCENTS IN STABILIZER ANALYSIS

2.1. Separation problems

2.1.1. Isolation of stabilizers and their decomposition products
The solution of so complex an analytical task as stabilizer analysis (see *Figure 1*) cannot be achieved without highly efficient separation processes for the comprehensive preliminary separation of the mixture of substances

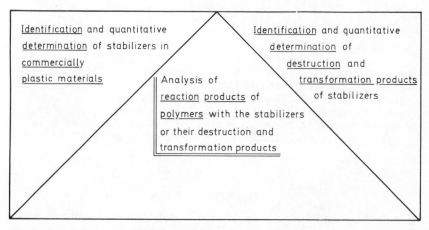

Figure 1. Problems of stabilizer analysis

despite all progress in the analytical chemistry, i.e. in particular by increasing the resolving power of absorption-spectrometric processes, by increasing the sensitivity in emission analysis as well as the selectivity of electrochemical methods, furthermore by chelatometry and also by radiochemical analysis. Therefore, separation problems, apart from structure research, are at the heart of modern scientific work and are to be dealt with also at the beginning of this paper. This, however, should not lead to the conclusion that efforts regarding *in situ* analysis are of secondary importance only, which is now as hitherto the method chosen for routine analysis of commercial products of comparable pre-treatment. Due to the large differences between molecule sizes of the predominantly low-molecular or oligomer stabilizers and the polymers and the great differences thus arising in the solubility, adsorption or diffusion behaviour, respectively, the separation routines even nowadays are subdivided into two partial processes, i.e. (a) separation of the low molecular substances from the polymer, and (b) separation of the mixture of low molecular substances. *Solid/liquid* extraction and *polymer precipitation* are even now the preferred methods amongst the numerous possibilities for the separation of low molecular substances. Taking into account the solvent effects on the extraction behaviour as well as the morphology of the polymers, optimum conditions of work have resulted from the abundance of publications for the appropriate polymer–stabilizer systems. An extract is given in *Table 1* (see refs. 1, 2).

Table 1. Extracting solvents for stabilizers

Polymer type	Substances extracted	Extracting solvents
PVC	Organo-tin stabilizers	Heptane–acetic acid 1:1
PVC	Organic-N-compounds stabilizers	Methanol or ether
PE	Antioxidants	Chloroform
POM	Phen. antioxidants	Chloroform
Rubber and	Stabilizers	Boiling acetone
vulcanizates	Accelerators	Boiling water

Apart from single-stage extraction, selective stage extractions are particularly favourable for the problems set by the analysis of plastics auxiliaries, because they already permit a preliminary separation of complex systems, although it is a rather coarse one. An excellent example for the selective extraction of stabilizers is amongst others that of the analysis of polyformaldehyde types[3] where the antioxidant used is extracted with chloroform and the heat stabilizer (dicyanodiamide) subsequently with methanol.

Under the new aspect of the stabilizer analytical techniques already described, however, even simple extraction processes present some complications which may result in faulty interpretation. From the function of the stabilizers to be expected, a high reactivity and lability result of necessity, which may lead to rearrangements, decompositions and similarly conditioned losses already with these first analytical operations. Despite the avoidable oxidative changes of the stabilizers, even the crushing phase of the polymer preceding the extraction means a source of danger. As is a well-known fact, all polymers of larger chain-lengths suffer from a degradation under mechanical stress which usually starts by a chain rupture in the centre of the macro-molecule and results in the formation of macro radicals. This degradation is accelerated at low temperatures and even occurs when cutting with a knife. By means of quantitative examinations, Pazonyi et al.[4] were able to find a linear relationship between the radical concentration (determined by the consumption of diphenyl-picryl-hydrazyl) and the surface (*Figure* 2) when cutting polyethylene and plasticized PVC. They could prove that the chemical bonds with polymers which to some extent are present in the elastic state are cut by mechanical forces with 100 per cent efficiency. In the presence of inhibitors reactions between macro radical and inhibitor may occur even in the absence of oxygen. As a consequence reaction products with the polymers themselves, but also deactivation products of the intermediately formed inhibitor radical are to be expected.

$$\sim \xrightarrow{\text{mech. eng.}} 2R^{\bullet} + 2I \longrightarrow 2R + 2I^{\bullet} \longrightarrow I—I \qquad (I)$$

Thus, for instance, Kenyon[5] also accepts an attachment of butyl residues of dibutyl-tin-diacetate to PVC radicals formed by u.v. radiation according to

$$R^{\bullet} + (C_4H_9)_2Sn(CH_3COO)_2 \rightarrow R—C_4H_9 + C_4H_9Sn(CH_3COO)_2 \qquad (II)$$

which has been confirmed by examination results obtained by Frye et al.[6] with ^{14}C butyl-labelled organotin compounds. In the presence of oxygen the reaction probability of the macro radicals with the inhibitor system is increased considerably mainly if phenolic or aminic antioxidants are present. Further information losses on reaction-induced decomposition products mainly occur when the extracts obtained are re-concentrated. Even the inherent volatility of some antioxidants—above all that of the phenols and aromatic amines (*Table* 2)—is so high that it provides the basis of a direct

Table 2. Volatility of antioxidants (see ref. 24)

Antioxidant	Vapour pressure [mm Hg]	Loss of weight %(150°C)
2,6-Di-t-butyl-p-cresol	22.15	100
2-Benzyl-6-t-butyl-p-cresol	1.83	100
2,2'-Methylene-bis-6-t-butyl-p-cresol	0.169	19...28
Diphenylamine	7.52	100
N-i-Propyl-N'-phenyl-p-phenylene-diamine	0.59	40...53
N-N'-Diphenyl-p-phenylenediamine	0.032	2...3

determination in the polymer by vacuum sublimation as suggested by Yushkevichyute[7] for the determination of antioxidants in PE and also can be applied to the explanation of the mechanism of action of phenolic and aminic antioxidants in PE and PP[8]. Thus a separation by distillation of the 2,6-di-t-butyl-4-methylphenol from its dimer deactivation product at 100°C was successful and provided evidence for the isomerization of the primarily formed phenoxy radicals to oxybenzyl radicals and their re-combination to dioxydiphenylethane:

236

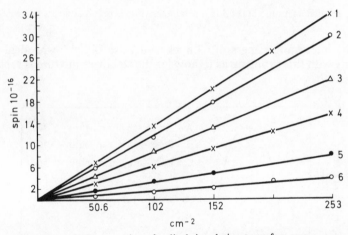

On the other hand the high volatility with the classical way of re-concentrating the extracts by distillation—even by evaporation of the solutions—of course will result in considerable loss of substance as we found in examinations with the same antioxidant[9]. We even found an evaporation loss of 0.75 per cent after 24 hours when storing Jonol with a large surface in stagnant air. This loss increased to 63 per cent when the chloroform solution was evaporated in a fume cupboard.

Figure 2. Consumption of radicals in relation to surface area

Distillation processes—even careful freeze-drying—should be avoided for quantitative work in systems of such high volatility as the stabilizers, especially if also quantity and type of the decomposition products are of interest. In such cases only an enrichment by chromatographic processes should be considered. From the point of view of separation from the polymer, gel-permeation chromatography is a possible method since the separation of molecules in the pores of the gels is mainly achieved according to particle size, and the large polymer molecules can be excluded from the separation process by suitable selection of the pore size distribution of the gels so that they leave the separation column together with the solvent. Such a separation of low molecular substances from polymer mixtures have already been described for plasticizers[12], oils[13] and methylsilanoles[14]. It is to be seen

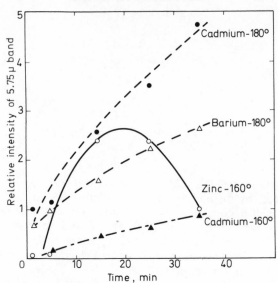

Figure 3. Relative intensities of the 5·75 µ band after various heat treatments in purified poly-
(vinyl chloride)

from the interesting papers of Čoupek and Pospíšil[10, 11], who deal syste-
matically with this problem, as to how far the stabilizer mixture is separated
subsequently.

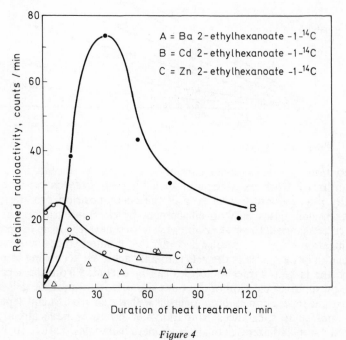

Figure 4

238

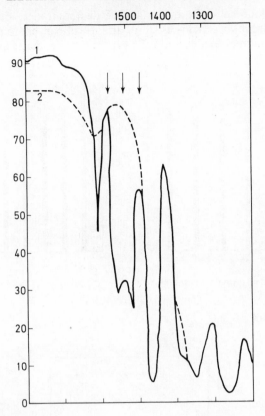

Figure 5. The i.r. spectrum of PVC with diphenylthiourea (1) before, (2) after heating at 165°C

With the well-known method of *precipitation of the polymer* from diluted solutions for the separation of soluble low molecular substances, the negative influences of the mechanochemical degradation during the crushing stage and the consequent reaction released are eliminated, but again this process needs a more thorough scientific consideration of the solution–precipitant effects on the stability, especially of the reaction products of stabilizers or their fragments with the polymer, under the new objectives of stabilizer analytical techniques. Such reaction products have been both determined and isolated with PVC, PE, PP and natural rubber.

Thus Frye *et al.* were able to provide evidence by i.r. spectroscopy[15] and radiochemistry[16] of the insertion of ester groupings from barium, cadmium or zinc-carboxylates in PVC after heat treatment (*Figures 3* and *4*). Tin contents and organic residues of organotin stabilizers in heat-treated PVC are indicated by the same authors[5, 17, 18]; Schlimper[19] and Hagen[20] proved the direct reaction between PVC and nitrogen-containing organic stabilizers (*Figure 5*) by elemental analytical examinations and i.r. and u.v. spectrometry. Phenolic antioxidants or their decomposition products in part were re-found in PP after oxidative degradation[8]; with rubber vulcanizates hydrochloric acid-resistant amine–rubber compounds are described after thermal oxidation in the presence of aromatic amines[21]. With

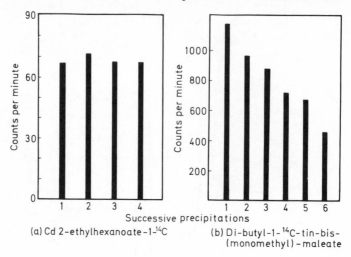

(a) Cd 2-ethylhexanoate-1-^{14}C

(b) Di-butyl-1-^{14}C-tin-bis-
(monomethyl)-maleate

Figure 6(a)   *Figure 6(b)*

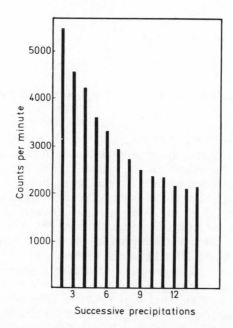

Successive precipitations

Figure 6(c). PVC with dibutyl-^{113}tin-bis-(monomethyl)-maleate after heat treatment

Figure 6. Retention of radioactivity by PVC after heat treatment: (a) Cd 2-ethylhexanoate-1-^{14}C;
(b) di-butyl-1-^{14}C-tin-bis-(monomethyl)-maleate

240

all this work, the polymer–stabilizer compounds are separated from the low
molecular products by precipitation processes and almost purified by
repeated 're-precipitation'. Whereas no structural modifications were
indicated in the course of repeated precipitations for PVC–N- and also
PVC–^{14}C-carboxylate compounds which were formed by reaction of PVC
with labelled cadmium soaps under the influence of heat [*Figure 6*(a)], a
decrease of the ^{14}C and ^{113}Sn contents in relation to the number of re-
precipitations of the appropriately labelled PVC-dibutyl-tin-bis-mono-
methyl-maleate compounds formed after heat treatment was to be noticed
[*Figures 6*(b) and (c)], which approaches a stable final value after ten to
twelve re-precipitations. The degree of the decrease of radioactivity and
thus of the chemical changes of the polymer–stabilizer compound both
depend on the solvent–precipitant couple and the quantities used (*Figure 7*)

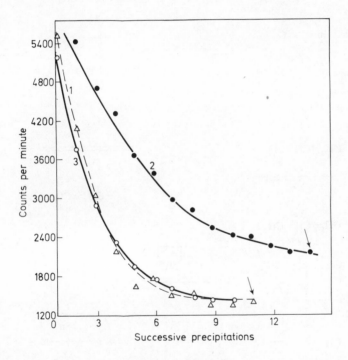

Figure 7. Influence of solvents on radioactivity on precipitation: (1) 3% PVC in THF; (2) 7.5%
PVC in THF; (3) 7.5% PVC in cyclohexanone

which are of decisive importance for stabilizer analysis, since it will be
necessary in future to refrain from the rule of thumb of a solvent–precipitant
relationship of 1:10 and to carry out special scientific preliminary tests.
As may be seen from *Figure 7*, the rearrangement rate of the polymer–
stabilizer compound with precipitation in about three per cent solutions in
THF with 2.5 times the quantity of methanol is higher (curve 1) as compared

to a 7.5% PVC–THF solution (curve 2). When using cyclohexanone (curve 3), the same results as with the more diluted PVC–THF solution were obtained. Frye et al.[17] concluded from the dependence on the tetrahydrofuran content on the one hand and the lability towards anhydrous hydrogen chloride on the other that a complex may exist between the corresponding tin compound and the polymer backbone molecule which is destroyed by tetrahydrofuran, cyclohexanone, methanol and mainly by hydrogen chloride as is to be seen from the illustration in *Figure 8*.

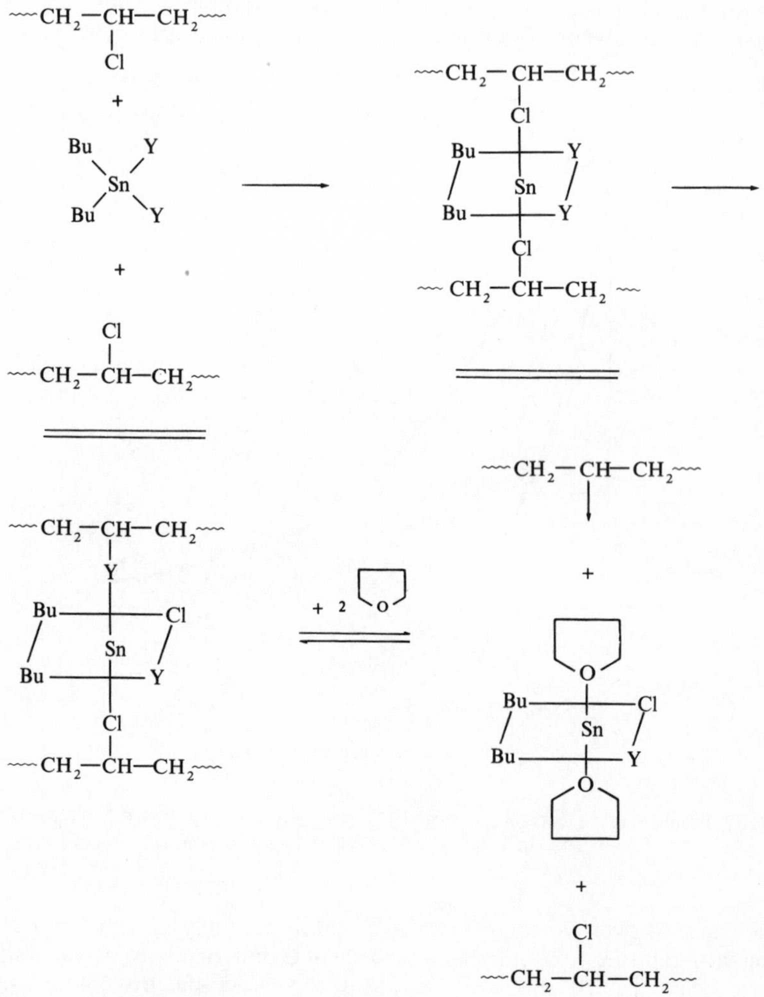

Figure 8

The rearrangement of the complex by the exchange of unstable chlorine atoms with the organic residue and its insertion after splitting of the complex is indicated by the remaining constant residual activities. The need for future definitive studies with regard to the interactions between solution–precipitant systems and polymer–stabilizer compounds may be deduced from the detail of this representation.

2.1.2. Separation of stabilizer systems

2.1.2.1. Thin-layer chromatography (TLC)—From all the known analytical separation processes, the greatest success so far for the problems of the stabilizer analysis has been obtained by thin-layer chromatography. Good separation efficiency in connection with high separation speed and great variability of the detection possibilities are the most important causes for this choice of treatment. Optimum working conditions for thin-layer chromatographic separation and identification have been published for all known stabilizer groups (see *Table 3*). Apart from the variation of carrier

Table 3. Thin-layer chromatographic methods for separation of stabilizers[1]

Substances separated	Stationary phase	Mobile phase	Detection
Organo-tin stabilizers	Silica-gel G	*n*-Butanol–acetic acid, 98:2	u.v. + Br_2-dithizone in chloroform
Organic stabilizers	Silica-gel G	Ethanol-free chloroform	
Antioxidants	Silica-gel G	1. Benzene 2. Light-petroleum– benzene– acetic acid = 1:1:1 3. Cyclohexane– diethylamine 4:1 and others	0.05% $Fe_2(SO_4)_3$ in sulphuric acid $+0.2\%$ $K_4Fe(CN)_6$ 1:1 0.1% 2,6-Dichloro-*p*-benzoquinone-4-chloramine in Ethanol
Salicylates Resorcinol-derivatives	Alumina + polyester $+H_2O + C_2H_5OH$	*m*-Xylene Formic acid	Diazotized 1-Amino-anthraquinone
Benzophenones	+ Sodiumdiethyl-di-thiocarbaminate Silica-gel G	98:2 Chloroform–*n*-Hexane 2:1	Diazotized sulphanilic-acid or other diazotized agents
Benzotriazoles	Silica-gel G	Benzene Benzene–light petroleum 7:3	u.v.-light

material, mobile phase and spray reagents, multi-stage processes here also offer prospective possibilities for the separation of complex stabilizer systems. The thin-layer chromatographic separation process for antioxidants by van der Neut[22] should be mentioned here as an excellent example, where the antioxidants are first decomposed on silica gel by benzene in six groups of increasing R_F values, and afterwards are separated selectively with another nine eluant systems and identified by variation of four detection substances[1,2].

Despite all efforts for improvement of the reproducibility of the R_F values no thorough changes are to be noticed for the quantitative stabilizer analysis, although favourable conditions are provided by the possible combination with physical measuring methods of high sensitivity such as fluorimetry (x-ray and u.v.). Apart from thin-layer chromatographic interfering effects in the main the high purity requirements regarding not only all chemicals and apparatus but also the stabilizers to be investigated are of detrimental influence since impurities may result in uncontrollable quenching effects.

2.1.2.2. Gas chromatography (GC)—As may be deduced from values of the inherent volatility of the stabilizers already given—in particular those of the antioxidants—some stabilizers may be separated directly by gas chromatography with maximum efficiency. In the majority of the cases known these conditions, however, are not given so that separation by gas chromatography is only possible after the stabilizers have been transferred into easily volatile compounds. This method is not very important because of the material losses often encountered with it, although the separation efficiency of GC has not yet been reached by any other chromatographic process.

2.1.2.3. Liquid chromatography (LC)—Due to theoretical investigation into the problem, liquid chromatography could approach the dissolving power of GC. Liquid chromatography has sustained tremendous development over the past ten years and is today the most universal chromatographic principle. It is a well-known fact that the dissolving power in chromatography in general depends on three factors: the number of theoretical plates N, the relative selectivity $\Delta K/K$ and the relationship between the migration rate of the corresponding zone and that of the liquid phase (R)

$$R_s = (N/16)^{\frac{1}{2}} (\Delta K/K)(1 - R)$$

(K is here a thermodynamic member and corresponds to the median distribution coefficient of two neighbouring coefficients.) Under the assumption of almost equal selectivities in the gas and liquid phases and a uniform R value of 0.5 with GC and LC which is common in practice, an approach of LC to the efficiency of GC is given as far as comparable N values or heights of equivalent theoretical plates (HETP) which are closely related to this are obtained. They may be almost realized by newly developed carrier materials.

The new high-efficiency carrier materials are mainly particles with a solid core and a thin porous envelope with particle diameters in the micro range. Majors[23] has tested the commercial carrier materials in an automatic high-pressure liquid chromatograph system with antioxidants. He has separated a four-component mixture of aromatic amines at HETP values of ≪1 mm with good reproducibility on Corasil I (glass bed with single layer of porous silica gel) loaded with β,β'-oxydipropionitrile within ten minutes (*Figure 9*). He also achieved separations of phenolic antioxidants as well as quantitative determinations in extracts of polyacetals the results of which are much superior to those of the i.r.-spectrometric determination due to the high resolution with liquid–solid column chromatography (LSCC). Further applications of this more recent variant of LC in stabilizer analysis give ground for optimism.

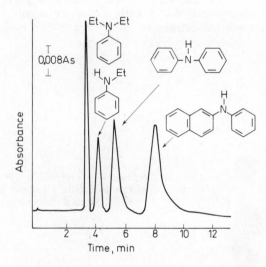

Figure 9. Separation of aromatic amine antioxidants using corasil 1

3. PROBLEMS OF STRUCTURE EXAMINATION

3.1. Stable decomposition products

Structure examinations of the stable decomposition and destruction products of stabilizers in principle correspond to classical organo-analytical concepts after separation from the polymers and exact preliminary separation. Therefore the methodological developments should not be described here in detail but progress made will be indicated by means of the present standard of knowledge. Up to now, analysis of the decomposition and transformation products has been done predominantly under the aspects of using stabilizer-containing polymers as packing materials for foodstuffs and semi-luxury goods as commodity, i.e. from the health point of view, and therefore it is essentially teamwork involving numerous disciplines with in part highly differentiated methodological solution possibilities. The most important of the decomposition products of the stabilizers were those of the tin compounds since they must be determinable even in the micro range because of some high toxicities.

Franzen *et al.*[25] were able to confirm the formation of tri- and mono-octyl-tin compounds, which may be formed by disproportionation, in pure di-*n*-octyl-tin-maleate after eight hours thermal load at 180°C by means of thin-layer chromatographic examinations which reveal with certainty alkyl-tin compounds in concentrations as low as 2γ tin:

$$2R_2SnY_2 \rightarrow R_3SnY + RSnY_3$$

245

The extent of disproportionation depends on the acid residue Y and is caused by strong Lewis acids as occur in the form of hydrogen chloride during thermal decomposition of PVC. Consequently, it was possible to identify these compounds even in heat-treated PVC films after aggravated extraction conditions (heptane–glacial acetic acid). From the negative result of the experiment after normal extraction (ether) it was an obvious conclusion

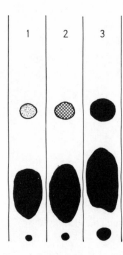

Figure 10. Detection of tri-*n*-octyl-tin in PVC after heat treatment: (1) di-*n*-octyl-tin-maleate; (2) extract before heat treatment; (3) extract after heat treatment

that here again a coordinative bonding similar to the PVC–organotin complexes already mentioned prevails (*Figure 10*). Therefore damage to health through the highly toxic tri-alkyltin compounds in heat-loaded PVC need not be expected because they are present only in small quantities (less than five per cent of the stabilizer used).

In addition, pure decomposition products were identified with stabilizers based on urea[20, 27]. Phenylisothiocyanate, diphenylurea, aniline and triphenylguanidine could be identified with certainty again by TLC when heating diphenylthiourea (15 minutes to 165°C) with 0.1N hydrochloric acid. The same decomposition products are also to be found in processed PVC films as well as nitrogen-containing PVC reaction products. The decomposition reactions resulting in these products are shown in *Figure 11*.

They can be detected not only in the i.r. but also in the u.v. spectrogram. Here the intensity maximum proper at $35\,500\ \mathrm{cm}^{-1}$ is much overlapped by decomposition products (*Figure 12*) and this forbids any direct determination in the films by u.v. spectrometry. This fact should hold true for numerous other systems as well so that direct determination may not always be fruitful.

The same decomposition products were found with the sulphur-free urea derivatives. In the course of PVC processing 2-phenyl-indole also suffers from decompositions which can be detected by the sudden extinguishing of luminescence. This extremely sensitive detection method is of general

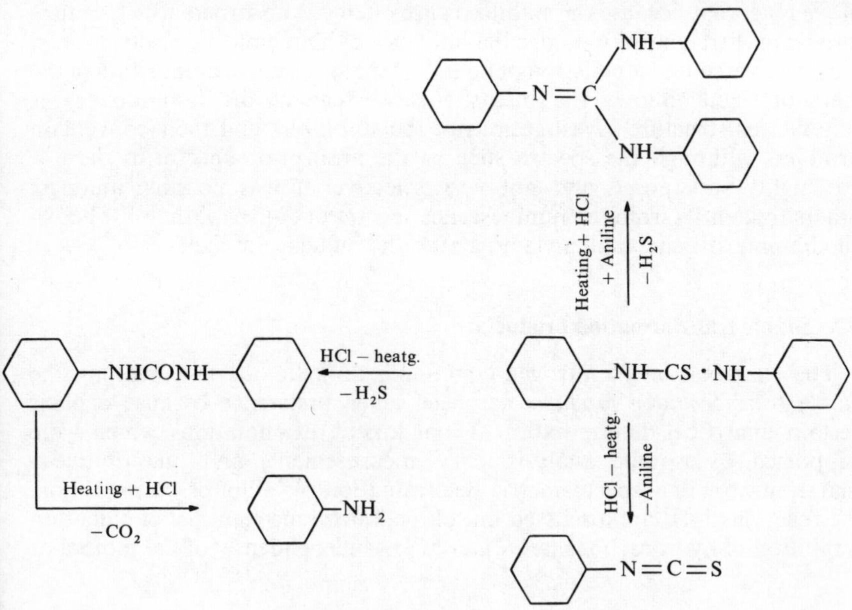

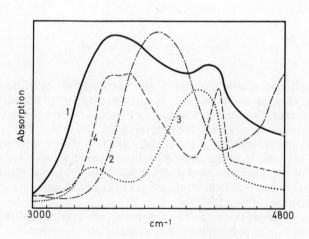

Figure 11. Destruction reactions of diphenylthiourea with hydrochloric acid after heating

Figure 12. The u.v. spectrum of: (1) diphenylthiourea; (2) diphenylurea; (3) aniline; (4) phenyliso-thiocyanate

importance for stabilizer analysis, in particular for following the time sequence of the reactions, because the stabilizers are often u.v. absorbers which change into an excited condition under the influence of light emitting electrons, and are deactivated under emission of light[26]. If the spectral decomposition of the emission light follows the intensity measurement, the luminometry is suitable for structure investigations of the stabilizers and their conversion products, although the spectra such as the absorption spectra in the u.v. are highly superposed and not very selective. It was possible amongst others to identify from the luminescence spectra of di-β-naphthyl-N-phenylenediamine, di- and triphenylamine after the influence of light.

3.2. Stable transformation products

The examination of nitrogen-containing antioxidants from among the stable *transformation products* has been given preference because of their certain analytical determination. From kinetic examinations which were supported by isotope analysis, e.p.r. measurements and quantitative—mainly u.v. or i.r.—spectrometric determination of inhibitor consumption, the reactions I–III are discussed out of the known mechanisms of oxidation inhibition of hydrocarbons (see *Figure 13*)—independently of the molecular

$$RO_2{}^{\cdot} + AH \longrightarrow ROOH + A^{\cdot} \tag{1}$$

$$RO_2{}^{\cdot} + \overset{|}{:}NH \longrightarrow RO_2{}^{:-} + {}^{\cdot}\overset{|}{N}H^+ \tag{2}$$

$$RO_2{}^{\cdot} + AH \rightleftharpoons (RO_2{}^{\cdot} \longleftarrow AH) \tag{3a}$$

$$(RO_2{}^{\cdot} \longleftarrow AH) + RO_2{}^{\cdot} \longrightarrow \text{stable products} \tag{3b}$$

$$RO_2{}^{\cdot} + AH \longrightarrow RO_2 - AH^{\cdot} \tag{4}$$

Figure 13. Reaction of antioxidants

size of the hydrocarbon. Amongst these reactions, that of hydrogen transfer for primary and secondary amines is perhaps the most probable one. Therefore amine radicals, radical ions, peroxy radical–inhibitor complexes as well as stable products are possible intermediary products. The deactivation process occurs mainly by way of dimerization, disproportionation or H transfer from the inhibitor radical. Quinone imides or diimides are final products of the last-mentioned reaction which were identified with certainty in the oxidation of N,N'-diphenyl-p-phenylene-diamine (*Figure 14*, 1) through their u.v. spectra.

With i.r. analysis of the extracts of oxidation products of rubber in the presence of N-phenyl-2-naphthylamine, the existence of dimer final products is deduced on account of the decrease of the NH band as well as the increase of the C—N and N—N bands.

These final products are formed by radical recombination[28] (*Figure 14*, 2). In addition to this, other reactions of the intermediary products with rubber can take place. These reactions were examined thoroughly by analyses with rubber vulcanizates by Tsurugi et al[21]. It was possible by means of acetone extraction of the oxidized vulcanizates which contained secondary mono- and diamines as well as tertiary amines as antioxidants, to isolate the amines not converted and the quinone diimines and to determine them quantitatively by a combination of potentiometric titration, GC and u.v. spectrometry. Semiquinone cations were found by treating the extraction residue with hydrochloric acid–ethanol and subsequent u.v. analysis. These semiquinone cations may only result from the hydrochloric acid splitting of the π-complex between a rubber–peroxy radical and the inhibitor formed according to reaction 3 (*Figure 14*). The extraction residue of the hydrochloric acid extraction also contained stable rubber–inhibitor reaction

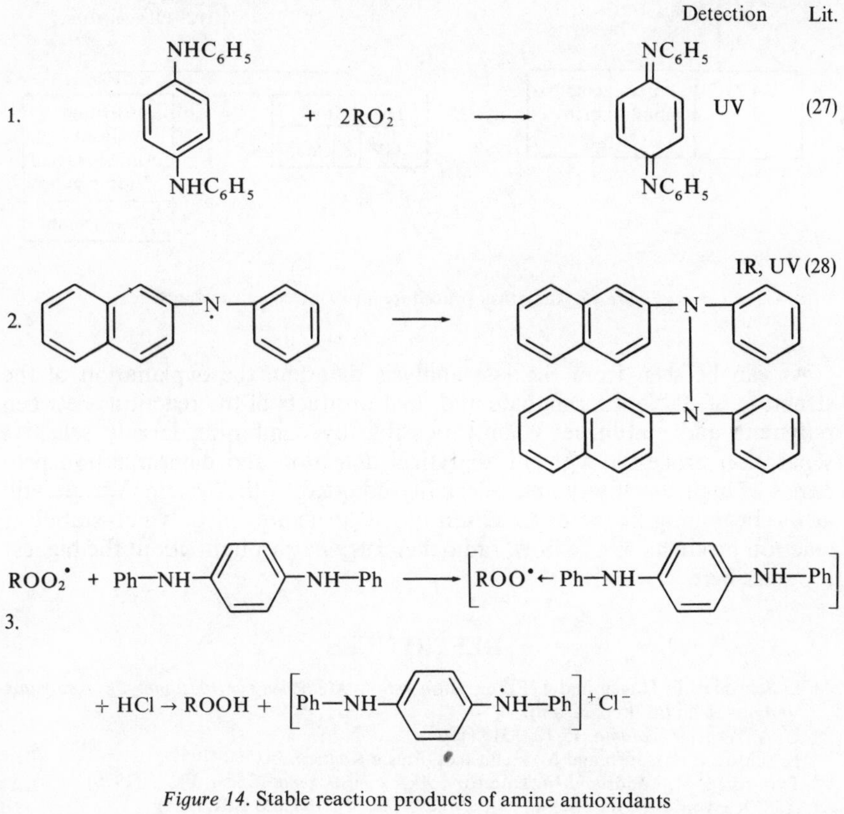

Figure 14. Stable reaction products of amine antioxidants

products. This successful combination of selective separation and analytical determination processes (*Figure 15*) permitted the direct identification, for the given example, of the reaction mechanism of inhibition during rubber oxidation through the intervention of the charge-transfer complex.

249

Depending on the resonance stabilization of the radicals which may be detected and interpreted in their structure, both benzenoid and quinonoid products are to be found with the phenolic antioxidants. The first stable conversion products of 4-methyl-phenoles found from systematic examinations with PP[8] were polymer quinones which change into low molecular quinones in the course of further oxidation.

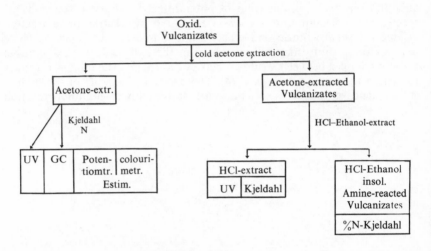

Figure 15. Analytical procedure for oxidized vulcanizates

As can be seen from the last analysis diagram, the explanation of the structure of stable intermediate and final products of the reactions between polymers and stabilizers is only possible by combining largely selective separation processes with all analytical detection and determination processes of high sensitivity and selectivity adapted to the system. We are still at the beginning in the explanation of the structures of polymer–stabilizer reaction products. Up to now, radiochemistry has brought about the biggest progress here.

REFERENCES

[1] E. Schröder, E. Hagen and J. Franz, *Ausgewählte Methoden zur Plastanalytik*, Akademie-Verlag: Berlin (in preparation).
[2] D. A. Wheeler, *Talanta*, **15**/12, 1315 (1968).
[3] E. Schröder, E. Hagen and M. Helmstedt, *Plaste Kautsch.* **14**, 560 (1967).
[4] T. Pazonyi, M. Tüdös and M. Dimitrov, *Angew. Makromol. Chem.* **10**, 75 (1970).
[5] A. S. Kenyon, *Nat. Bur. Stand. Circ. No. 525*, 81, 87, 100 and 106 (1953).
[6] A. H. Frye, R. W. Horst and A. Mako, *J. Polym. Sci.* A2, 1765 (1964).
[7] S. S. Yushkevichyute and Yu. A. Shlyapnikow, *Plast. Massy*, **12**, 62 (1966).
[8] L. L. Yasina, B. A. Gromov, V. B. Miller and Yu. A. Shlyapnikow, *Vysokomol. Soedin.* **8**, 1411 (1966).
[9] E. Schröder, and G. Rudolph, *Plaste Kautsch.* **10**, 22 (1963).
[10] J. Čoupek, S. Pokorný and J. Pospíšil, Lecture XI. Microsymposium on Macromolecules, Prague (1972).

[11] J. Protivová, J. Pospíšil and L. Zikmund, Lecture XI. Microsymposium on Macromolecules, Prague (1972).
[12] D. F. Alliet and J. M. Pacco, *Separation Sci.* **6**, 153 (1971).
[13] R. D. Mate and H. S. Lundstrom, *J. Polym. Sci.*, Part C, **21**, 317 (1968).
[14] F. N. Larsen, *Amer. Lab.* **10**, 10 (1969).
[15] A. H. Frye and R. W. Horst, *J. Polym. Sci.* **40**, 419 (1959).
[16] A. H. Frye and R. W. Horst, *J. Polym. Sci.* **45**, 1 (1960).
[17] A. H. Frye, R. W. Horst and M. A. Paliobagis, *J. Polym. Sci. A*, 1785 (1964).
[18] A. H. Frye, R. W. Horst and M. A. Paliobagis, *J. Polym. Sci.* **A2**, 1801 (1964).
[19] R. Schlimper, *Plaste Kautsch.* **10**, 19 (1963).
[20] E. Schröder, E. Hagen and S. Frimel, *Plaste Kautsch.* **14**, 158, 814 (1967).
[21] J. Tsurugi, S. Murakami and K. Goda, *Rubber Chem. Technol.* **44**, 857 (1971).
[22] J. H. van der Neut and A. C. Maagdenburg, *Plastics, London*, **31**, 66 (1966).
[23] R. E. Majors, *J. Chromatogr. Sci.* **8**, 338 (1970).
[24] K. Thinius, *Stabilisierung und Alterung von Plastwerkstoffen*, Akademie-Verlag: Berlin (1969).
[25] V. Franzen and G. Neubert, *Chemiker Ztg.* **23**, 801 (1965).
[26] R. N. Nurmukhametov, L. V. Bondareva, D. N. Shigorin, N. V. Mikhailov and L. G. Tokareva. *Hochm. Verb. UdSSR*, **6**, 1411 (1964).
[27] C. E. Boozer, G. S. Hammond, C. E. Hamilton and N. S. Iyotirindra. *J. Amer. Chem. Soc.* **77**, 3233 (1955).
[28] J. G. Angert and A. S. Kuzminski. *J. Polym. Sci.* **32**, 1 (1958).

PROBLEMS INCURRED IN STABILIZING POLYMERS AGAINST OXIDATIVE DEGRADATION. MODEL STUDIES ON ACTIVE FUNCTIONAL GROUPS IN THE POLYMER SUBSTRATE AND NEW INHIBITORS

HERBERT NAARMANN

6700 Ludwigshafen/Rhein, BASF/WK, F 206, West Germany

ABSTRACT

An investigation was made into the formation of functional groups that occur in unstabilized polyethylene during oxidation under the conditions of the induction test. Model substances were synthesized in which the nature and amount of functional groups, e.g. peroxide, carbonyl, hydroxyl, vinyl, etc., differed. The behaviour of these specimens was studied under the conditions of the oxidation test. The relationship of the induction time to the type of functional group allowed a sequence of reactivity to be drawn up, i.e.

$$-OOH > CO > -CH_2CH=CH_2 > -O-CH=CH_2 > -C=C-$$

$$> -\overset{\overset{\displaystyle CH_3}{|}}{C}-H > CHOH > CH-OR$$

Here the damaging effect of the peroxide group on the alkyl ether group decreases from left to right. Comparison was made with a specimen stabilized with substituted phenols and a polymethylene produced by the decomposition of diazomethane.

The second part of the paper presents substances of the following structures containing —S—:

$$CH_2=CH-S-CH_2-CH_2-R \qquad CH_2=CH-\overset{\overset{\displaystyle \oplus}{|}}{\underset{\underset{\displaystyle R_1}{|}}{S}}-CH_2-CH_2-OH \; X^{\ominus}$$

$$HO-CH_2-CH_2-S-CH_2-CH_2-S-CH_2-CH_2-OH$$

The functional groups of these compounds compensate the injurious effect of the groups in the substrate, with the result that the compounds act as inhibitors. For example, the thioether bridges are converted to $>SO$ and $>SO_2$ derivatives. The sulphurous substances that also contain a vinyl double bond allow capture by disproportionation and addition reactions.

Although much work has been devoted to it, the problem of stabilizing polymers against oxidative degradation is still widely discussed. It retains its importance for the following reasons. (1) Increasing demands are being imposed by plastics processors and end users. (2) There are no technically

253

perfect products available, i.e. most products display some defects, have fluctuations in their molar weights and morphology, and contain some impurities. Improvements in manufacturing processes and changes in the specifications for the raw materials may appear to be insignificant microscopically but, when regarded in detail, are often the reason for troublesome secondary effects. The need for developing even more effective inhibitors has not lost its urgency.

Because of this, oxidative reactions in polymers assume great practical importance. Much work has been carried out in this field. It is assumed that the first stage of oxidation is the formation of radicals and of hydroperoxides, which have been detected in all polyolefin oxidation processes.

We have tried to establish a relationship between oxidative degradation and the chemical structure of products with a defined amount of functional groups in the light of the many hypotheses that have been put forward and the analytical data available. Polyethylene was taken as a model.

Table 1

Character	Low-density polyethylene	High-density polyethylene
Density (g/cm^3)	0.918–0.920	0.925–0.950
Melt flow index MFI 190/2/16 g/10 min	1.2–1.7	0.1–0.6
CH$_3$/1000C	30–36	1
Crystallinity	34–40	75–80
Vinyl	0.05–0.2	1.0–0.3
—CH=CH$_2$		
Vinylidene	0.5–0.8	0.05–0.2
>C=CH$_2$		
1,4-*trans* —C=C—	—	—
>CO	0.01	0.01
—OH	+	+

It is generally realized that conventional polyethylenes are not uniform polymers consisting of nCH$_2$—CH$_2$ sequences (cf. *Table 1*). Strictly speaking, they must be considered as statistical copolymers with CH$_3$, vinyl, vinylidene and hydroxyl side groups and with carbonyl groups and double bonds in the chain. If the specimens have aged, hydroperoxide groups can also be detected, although they are only present in traces.

The induction period method was adopted to measure the oxidative degradation of the polymers. The polethylene specimens were flushed at 180°C with preheated pure oxygen, and the oxygen uptake was measured as a function of time (cf. *Figure 1*).

From *Figure 2*, it can be seen that the induction time for low-pressure polyethylene is different to that for high-pressure polyethylene. Hence, the decided differences in oxidation stability must be due to the structure.

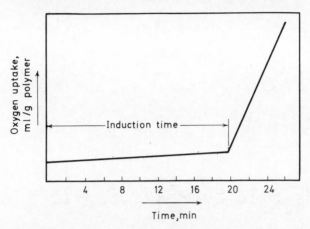

Figure 1. Oxygen uptake at 180°C as a function of time. The flat portion represents the induction time.

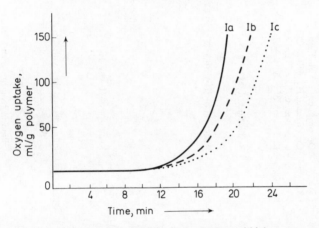

Figure 2. Induction times measured at 180°C for low-pressure and high-pressure polyethylene. I ——— high pressure polyethylene; Ib --- low-pressure polyethylene; Ic ··· Ib reprecipitated.

In the case of the low-pressure type, it appears that traces of heavy metals arising from the polymerization process may also be of some significance: a decidedly higher stability is observed after precipitating in the presence of complexing agents (cf. *Figure 2*, curve Ic).

In order to determine which functional group is responsible for the degradation mechanism model compounds with the structures listed in *Table 2* were synthesized and tested.

The effect on the oxidation stability of increasing the proportion of the functional group is shown in *Figure 3*.

A good idea of which functional groups are responsible for oxidative degradation under the test conditions can be obtained from *Figure 4*. Each carbon skeleton was allotted the same amount of functional groups in

255

Table 2. Synthesized model compounds. Structure, melt flow index, method of synthesis, and details of functional group

No.	Structure	MFI	Method of synthesis	Functional group type	%
Ia	$(CH_2-CH_2)_n$	3.8	Decomp. of diazomethane	—	—
II	$(CH_2-CH_2)_{90}CO-$	1.8	Copolym. $CH_2=CH_2/CO$	CO	1.1
III	$-(CH_2-CH_2)_{10}-CO-$	4.0	Copolym. $CH_2=CH_2/CO$	CO	8.5
IV	$-(CH_2-CH_2)_{90}-CH-$ | OH	1.6	Reduction of II with $NaBH_4$	$-CH-$ | OH	~0.6
V	$-(CH_2-CH_2)_{10}-CH-$ | OH	3.2	Reduction of III with $NaBH_4$	$-CH-$ | OH	~5.5
VI	$-(CH_2-CH_2)_{10}-C-$ | $=O$ | $HOOH$	*	Reaction of III with H_2O_2 50°C, H^+	$-C-$ | $=O$; $HOOH$; $-C-OH$	~9
VIa	$-(CH_2-CH_2)_{90}-C=O$ | $HOOH$	*	Reaction of II with H_2O_2 50°C, H^+	$-C-$ | $=O$; $HOOH$; $-C-OH$	~1.2
VII	$-(CH_2-CH_2)_{10}-CH-$ | CH_3 | OOH	2.1	Ullmann reaction with CH_3Cl	$-CH-$ | CH_3 ; $-C-OOH$	~5.0
VIII	$(-CH_2-CH_2)_{12}-C-$ | CH_3 | CH_3	*	Ullmann reaction with CH_3Cl followed by (a) Admission of air (b) Reaction with H_2O_2	$-C-$ | CH_3 ; $-C-OOH$	~8.0

No.	Structure	MFI	Method of synthesis	Functional group type	%
IX	$(CH_2-CH_2)_{10}-CH-$; $O-CH_3$	2.9	Reaction of V with diazomethane	$-O-CH_3$	~9
IXa	$(CH_2-CH_2)_{90}-CH-$; $O-CH_3$	1.2	Reaction of IV with diazomethane	$-O-CH_3$	~1
X	$(CH_2-CH_2)_{10}-CH-$; $O-CH=CH_2$	†	Vinylation of V with $HC\equiv CH$	$-O-CH=CH_2$	~12
Xa	$(-CH_2-CH_2)_{90}-CH-$; $O-CH=CH_2$	1.5	Vinylation of IV with $HC\equiv CH$	$-O-CH=CH_2$	~1.5
XI	$-(CH_2-CH_2)_{10}-C-$; OH ; $CH=CH_2$	†	Ethynylation of III with $HC\equiv CH$, and following reduction	$OH-C-CH=CH_2$	~8
XIa	$-(CH_2-CH_2)_{90}-C-$; OH ; $CH=CH_2$	1.6	Ethynylation of II with $HC\equiv CH$ and following reduction	$OH-C-CH=CH_2$	~1
XII	$-(CH_2-CH_2)_9-C=C-$	†	Dehydration of V	$C=C$	~7
XIIa	$(-CH_2-CH_2)_{89}-C=C-$	1.4	Dehydration of IV	$C=C$	~0.8

* Decomposes
† Crosslinks

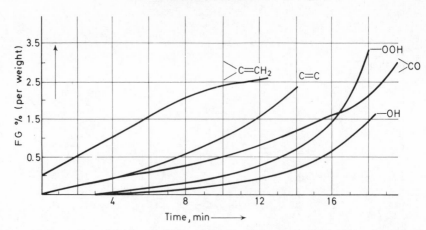

Figure 3. Proportion of the functional groups in polyethylene I (unstabilized) as a function of time at 180°C in the induction period test.

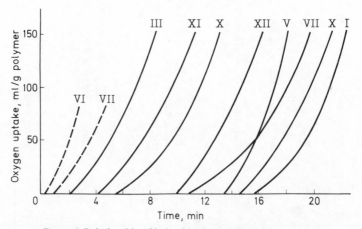

Figure 4. Relationship of induction time to the polymer structure.

order to facilitate comparison; in this case, one functional group on an average was allotted to each ten CH_2—CH_2 groups. The following conclusions can be derived by studying the relationships thus obtained between structure, i.e. functional groups, and induction time.

(1) The induction time is directly proportional to the stability of the product concerned.
(2) Peroxide groups are the most active and are primarily responsible for oxidative degradation (cf. *Figures 4* and *5*). The degradation is so rapid that the specimen commences to decompose when it is being heated up. From about 80°C onwards, i.e. about 100°C below standard, measurement is no longer possible.
(3) Vinyl groups in the sidechain induce degradation more strongly than —C=C— groups in the chain.

258

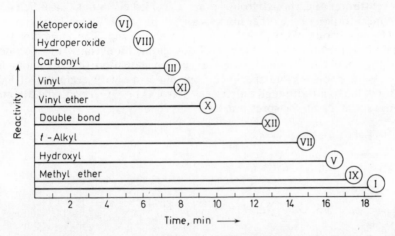

Figure 5. Relationship between chemical structure and induction time. The Roman figures in the circles are the serial numbers of the products listed in *Table 2*. I denotes a general-purpose low-density polyethylene

(4) Methylated hydroxyl groups have practically no effect.
(5) An ideal —CH$_2$—CH$_2$— polymer (Type Ia) without any side groups whatsoever is much superior to the general-purpose polymer I (high-pressure polyethylene), cf. *Table 3*.

The behaviour of this 'ideal polymer' will be considered later in the section dealing with the effect of antioxidants and in the part where their behaviour is compared to that of general-purpose polyolefins stabilized with conventional additives.

The relationship between the concentration of the functional groups and their efficiency is explained below.

In *Table 3*, the induction times for products with one functional group distributed over ten CH$_2$—CH$_2$ units are compared to those for one functional group distributed over ninety CH$_2$—CH$_2$ units. (The ratios in each case are statistical means.) The concentration of the functional groups concerned can be derived from *Table 2*. It can be seen that differences in

Table 3. The relationship of induction time to the concentration of the functional groups (FG)

Type and No.	Ratio 1 FG to 10 —CH$_2$—CH$_2$— Induction time, min	No.	Ratio 1 FG to 90 —CH$_2$—CH$_2$— Induction time, min
Ketoperoxide VI	1	VIa	*ca.* 1
Hydroperoxide	2	—	—
Carbonyl III	5	II	11.5
Vinyl + hydroxyl VI	7.5	XIa	10
Vinyl ether X	9	Xa	12
C=C in the chain XII	13	XIIa	15
t-Alkyl VII	14.5	—	—
Hydroxyl V	16	IV	18
Methyl ether IX	17	IXa	19

concentration of carbonyl groups exert a decided effect but that differences in concentration of the other functional groups are only slightly gradated.

The behaviour of specimen Ia is quite interesting. The induction time fluctuates in relationship to the time of storage and the ageing of the specimen. Freshly prepared polymethylene (Ia) has an induction time of about 65, whereas a one-week-old specimen kept under air with exclusion of light has a value of 40, although only traces of —CO or —OH could be detected in the specimen by the spectrometer.

Specimens for comparison purposes:

High-pressure polyethylene I	15–20
Ideal polymethylene Ia (—CH$_2$—)$_n$	40–65

I + 0.1 % wt   CH$_3$—⬡(t-C$_4$H$_9$)(t-C$_4$H$_9$)—OH‡   45–50

The data submitted for comparison purposes in *Figure 4* and *Table 3* are the induction times in which 100 ml of oxygen were consumed for each gramme of resin. A clear gradation is evident on comparing the efficiency.

The results of the oxidation stability tests shown in *Figure 4* arrange the functional groups in a sequence ranging from —OO— groups to the general-purpose type I (high-pressure) polyethylene.

It is surprising to note that CO groups favour oxidative degradation much more than do —C═C— groups. A tertiary hydroxy group combined with a vinyl group corresponding to that in compound XI is more active than a vinyl ether group (that in compound X) or a double bond in the chain (compound XII). This fact can be ascribed to the preference of hydroperoxide formation in compound XI.

The results presented in *Figures 4* and *5* allow the compound to be arranged in the sequence of oxidative degradation. The question thus arises whether other factors, such as contamination of the specimens by traces of heavy metals or other substances, may activate or inactivate oxidative degradation. It was already demonstrated in *Figure 2* that a precipitated specimen of low-pressure polyethylene (Ic) had a higher induction time than the original material (Ib).

In none of the cases investigated, could any change in the reactivity sequence presented in *Figure 4* be observed by varying the test conditions for the determination of induction time, i.e. by exposing the specimens to temperatures of 150°C and 100°C instead of to the temperature laid down, viz. 180°C.

The gradation in reactivity and the effect of the functional groups concerned, as determined on the model substances, can be exploited in practice.

(1) The injurious functional groups in polymers may be eliminated or converted into others that are less reactive.

---

‡ 2,6-Di-*t*-butyl-4-methyl-phenol.

(2) The functional groups can be rendered ineffective by the introduction of competitive free radicals to inhibit oxidation of the polymer.

In all radical-induced autoxidation processes, substitution reactions take place. some of which are self-catalysing, and the dioxygen reacts as a di-radical with two unpaired electrons $\cdot O{-}O\cdot$. Consequently, another means of capturing less reactive radicals is the addition or formation of other radicals.

Radicals are formed by mechanochemical reactions occurring during the processing of polymers, e.g. during calandering or extrusion, but this is beyond the scope of the present paper.

MEASURES FOR PREVENTING OXIDATIVE DEGRADATION

(1) Chemical reactions with the polymers.
(2) The use of additives.

Tests on the model substances have demonstrated that peroxy groups are responsible for the most severe degradation. Therefore the following measures are required:

(a) The formation of —OO— groups should be prevented from the very beginning (but this is not possible in practice).
(b) Existing —OO— groups should be removed.

Peroxide decomposition

The stability of peroxides depends on their structure, and on the ambient conditions including temperature.

Much work has been done in relating differences in peroxy group stability to the structure[1]. Materials have to be found which will decompose all kinds of —OO— groups that have formed in the polymer.

Radical-induced decomposition

Morse[2] found that alkyl radicals derived from a solvent induce the decomposition of hydroperoxides in the presence of oxygen. The oxygen and the hydroperoxide compete for solvent radicals.

Thermally-induced decomposition

This also involves radical-induced decomposition. Kharasch et al.[3] studied some model reactions and proposed that, in fact, reduction takes place.

Solvent effects

Thomas et al.[4] studied decomposition in a range of solvents and suggested that the effect observed was induced decomposition of the peroxide by the solvent.

Metal ion effect

Metals or their ions, e.g. copper, manganese and cobalt, may act as catalysts and speed up decomposition. They are highly efficient autoxidation

261

catalysts. Dean[5] postulated reversible formation of a metal–peroxide coordination complex following an electron transfer. Kharasch *et al.*[3] proposed that the relative oxidation of alkylperoxy radicals is the source of oxygen.

Reduction methods

Peroxides can be reduced according to equation 1.

$$R'H + ROOH \rightarrow \begin{bmatrix} RO^{\bullet} \\ {}^{\bullet}R' \end{bmatrix} + H_2O \qquad (1)$$

Examples of reducing agents (R'H) that can be used for this purpose are –S– compounds, P(III) derivatives, amines, sugars and all compounds with an unstable hydrogen atom[6]. In all cases free radicals are formed.

Non-radical-induced decomposition

The anionic mechanisms responsible for non-radical-induced decomposition are classified into the following three groups[7].

Catalysis by acids

Solvents with a high dielectric constant favour the decomposition reaction. The stronger the acid, the greater the rate of decomposition[8].

Catalysis by bases

Even weak bases or dinitriles, e.g. phthalonitrile, are powerful reagents for promoting decomposition[3, 9].

Intermolecular rearrangement

In this case, the end products are formed via an intermediate without the formation of radicals.

Stoichiometric reactions

This important type of reaction closely resembles the addition of reducing agents. Examples are the reaction between hydroperoxides and olefins yielding epoxides[10] and the reactions between hydroperoxides and sulphides[11].

MODEL REACTION TO ELIMINATE ACTIVE FUNCTIONAL GROUPS

The induction time of model compounds I and II can be lengthened by reducing them with a highly reactive agent[6].

Another means of lengthening the induction time is to add sulphur or carbon black.

We tried to exploit the knowledge we had gained from our model experiments in finding highly reactive material that would either block the functional groups or would compete with them for oxygen.

The inhibitors commonly used are phenols and aromatic amines, which terminate the kinetic chain. The concentration of inhibitor undergoing oxidation is normally about one per cent or, in some cases, less.

Table 4. Induction times of polyethylene containing reducing agents

No.	Type	Specification	Induction time, min
I	High-pressure polyethylene	*Table 1*	15–20
Ib	Low-pressure polyethylene	*Table 1*	∼20
Ia	Polymethylene	No funct. groups	∼60
XIII	I reduced by SO_2	Peroxides	∼25
XIV	I reduced by H_2	or CO	∼35
XV	I reduced by H_2S	undetectable	∼40
XVI	I reduced by $NaBH_4$	by i.r.	∼45

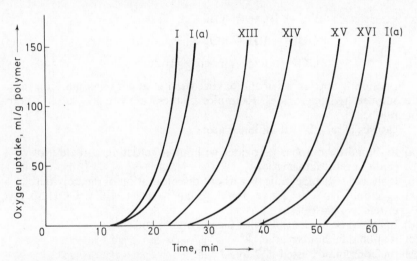

| I | stabilized with CH_3—⟨ring with t-C_4H_9, t-C_4H_9 ⟩—OH | | 45–50 |
| | 0.1 % wt of | | |

*2,6-Di-t-butyl-4-methyl-phenol.

Oxygen uptake, ml/g polymer

I I(a)   XIII   XIV   XV XVI I(a)

Time, min ⟶

Figure 6. Induction times of modified samples compared to that of the basic products.

Should the chain terminating reaction compete successfully with the propagation reaction,

$$RO_2{}^{\cdot} + RH \rightarrow ROOH + R^{\cdot} \qquad (2)$$

where RH represents a segment of the polymer, then the inhibitor must have one or more labile H atoms. As a result of the H atoms in the molecule, the inhibitor reacts more readily with the oxygen than the functional groups in the polymer.

$$IH + O_2 \rightarrow I^{\cdot} + HOO^{\cdot} \qquad (3)$$

where IH represents the inhibitor. This reaction leads to an active free radical.

263

$$HOO^{\bullet} + RH \rightarrow HOOH + R^{\bullet} \qquad (4)$$

(A) If the number of free radicals formed during the oxidative degradation of one ROOH is greater than unity, each chain scission will be accompanied by the formation of more than one active free radical. Thus at any inhibitor concentration, oxidation will proceed autocatalytically during the induction period.

(B) If the number of kinetic chains formed by the decomposition of one ROOH is less than unity, the inhibitor is oxidized directly, and the reaction proceeds by the mechanism shown below, which has been proposed by Shlyapnikov and Miller[14].

$$IH + O_2 \rightarrow I^{\bullet} + HOO^{\bullet} \qquad (3)$$

$$I^{\bullet} + HOO^{\bullet} + RH \rightarrow R^{\bullet} + \text{inactive products} \qquad (4)$$

$$R^{\bullet} + O_2 \rightarrow RO_2^{\bullet} \qquad (5)$$

$$RO_2^{\bullet} + RH \rightarrow ROOH + R^{\bullet} \qquad (6)$$

$$RO_2^{\bullet} + IH \rightarrow ROOH + I^{\bullet} \qquad (7)$$

$$ROOH + RH \rightarrow RO_2^{\bullet} + \text{inactive products} \qquad (8)$$

$$ROOH + R_2S \rightarrow \text{inactive products} \qquad (9)$$

Equation 8 applies when the oxidized material contains inhibitors that decompose hydroperoxides. Examples of these are organic sulphur compounds.

Features of an efficient inhibitor are:

(a) It should decompose peroxides without re-initiating autoxidation or a chain of radical formation;
(b) It should scavenge radicals without discolouration of the polymer;
(c) It should have an active double bond to react with ROOH to give unreactive epoxides;
(d) It should be compatible with the polymers; and
(e) It should be non-volatile.

All the compounds listed in *Table 5* can:
(a) Decompose hydroperoxides;
(b) React with ROOH via vinyl groups forming epoxides†;
(c) React with oxygen; and
(d) Scavenge radicals

$$\underset{A}{CH_2{=}CH}{-}\underset{B}{S}{-}\underset{}{CH_2{-}CH_2}{-}\underset{C}{R}$$

All these model compounds have three different units.

(A) $CH_2{=}CH{-}$ (a) radical absorbing, polymerization
(b) O_2 reaction $\rightarrow$ epoxides (hydroperoxide)
(c) $-OH$ addition $\rightarrow$ ethers

---

†Berlin[15] studied the effect of double bond systems as inhibitors in thermo-oxidative processes.

Table 5. Model compounds for stabilization

First series

No.	CH$_2$=CH—S—CH$_2$—CH$_2$—R	B. pt, °C
XVII	X—OOC—CH$_2$—CH$_2$—N⟨ ⟩O	160 mm Hg
XVIII	X—OOC—CH$_2$—CH$_2$—N⟨ ⟩N—CH$_3$	184–186 mm Hg
XIX	X—OOC—CH$_2$—CH$_2$—N⟨ ⟩N—CH$_2$—CH$_2$—COO—X	F. pt 35–40
XX	X—OOC—CH$_2$—CH$_2$—NH—(CH$_2$)$_3$—N⟨ H ⟩	195 mm Hg
XXI	X—O—CH$_2$—CH$_2$—CN	120–124 mm Hg
XXII	(X—O)$_3$—P	112–115/0·2 mm Hg

X   CH$_2$=CH—S—CH$_2$—CH$_2$

(B)   —S—   (a) O$_2$ reaction → —S— sulphoxides
                                           ‖
                                           O

                        ⟩SO$_2$ sulphones
                   In the latter case, this entailed that about 20 litres
                   of oxygen are absorbed by 1 mol of compound
                   XVII, XVIII, XX, XXI
                   by $\frac{1}{2}$ mol of compound XIX
                   by $\frac{1}{3}$ mol of compound XXII (if only S reacts) and
                   by $\frac{1}{4}$ mol of compound XXII (if S and P react).
                   (b) radical scavenging—chain transfer.
(C)   R—   (a) Contains reducing and radical-scavenging groups
                   (except XXI). XXI decomposes hydroperoxides in
                   a base-catalysed reaction via the —CN groups.

Bolland[13] studied the autoxidation and autoinhibition of saturated thioethers and allylic sulphides in the presence of radical generators. He observed that the addition of a fresh quantity of a radical generator restored the rate of oxidation. This phenomenon indicates that efficient inhibitors are formed during the autoxidation process. These compounds are taken as monomers in various copolymerization reactions, in which case they act as chain transfer agents, i.e. they avoid crosslinkage during polymerization, an important function in the polymerization of dienes.

The efficiency of compounds XVII to XXII was investigated by adding them to general-purpose polyethylenes I and Ib and to the model polymers VI, VIa and VIII, which contain —O—O— groups.

In all cases, polyethylene I was inhibited with 0.1% wt of compounds XVII to XXII. The induction time curve for the uninhibited sample (I) is shown on the left of *Figure 7*, and the corresponding curves for the six inhibited samples on the right. The improvement in stability achieved by

265

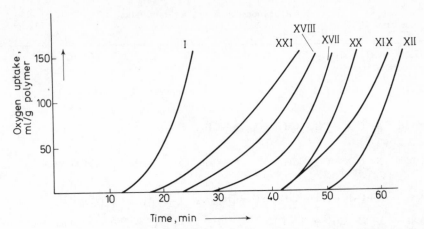

Figure 7. Induction times for general-purpose unstabilized polyethylene and for polyethylene stabilized with inhibitors Nos. XVII to XXII.

Table 6. Correlation between induction time and sample VI, which contains —OOH groups, after stabilization

No.			Induction time, min
VI Ketoperoxide	content of $-\overset{\overset{\text{O}}{\|}}{\text{C}}-$ HOOH	9%	1
	+ XVII 0.1% wt		6
	XVIII 0.1% wt		5
	XIX 0.1% wt		8
	XX 0.1% wt		8
	XXI 0.1% wt		3
	XXII 0.1% wt		10
VIa Ketoperoxide	content of $-\overset{\overset{\text{O}}{\|}}{\text{C}}-$ HOOH	1.2%	1
	XVII 0.1% wt		6
	XVIII 0.1% wt		8
	XIX 0.1% wt		8
	XX 0.1% wt		10
	XXII 0.1% wt		15
VIII Hydroperoxide	content of $-\overset{\overset{\text{CH}_3}{\|}}{\underset{\|}{\text{C}}}-$ OOH	~8%	2
	XVII 0.1% wt		5
	XVIII 0.1% wt		6
	XIX 0.1% wt		8
	XX 0.1% wt		8
	XXI 0.1% wt		5
	XXII 0.1% wt		12

the inhibitors containing sulphur, i.e. XVII to XXII, can be seen clearly from *Figure 7*. It now remains to be clarified whether these inhibitors are still efficient if they are added direct to polymers containing —OOH groups, e.g. the model substances VI, VIa and VIII.

From *Table 6* it can be seen that the induction time can be increased by a factor of between three and ten.

The stabilizing agents in the second series have the formula

$$HO—CH_2—CH_2—S—CH_2—CH_2—S—CH_2—CH_2—OH$$
$$R_1$$

These compounds were synthesized by adding mercaptoethanol (XXIII) to vinyl hydroxyethyl ether (XXIV).

$$HO—CH_2—CH_2—SH + CH_2{=}CH—S—CH_2—CH_2—OH$$
$$\text{XXIII} \qquad\qquad\qquad \text{XXIV}$$

$$\rightarrow HO—CH_2—CH_2—S—CH_2—CH_2—S—CH_2—CH_2—OH$$
$$\text{XXV}$$

Compound XXV is monoesterified with H_3PO_3 to give compound XXVI or diesterified with H_3PO_3 to give compound XXVII. Likewise, compounds XXVIII and XXIX are obtained by monocyanoethylation and dicyanoethylation of compound XXV.

Table 7

No.		Melting point, °C
XXV	$HO—CH_2—CH_2—S—CH_2—CH_2—S—CH_2—CH_2—OH$	122
XXVI	$(HO—CH_2—CH_2—S—CH_2—CH_2—S—CH_2—CH_2—O)_3P$	79–82
XXVII	$(O—CH_2—CH_2—S—CH_2—CH_2—S—CH_2—CH_2—O)_6P_2$	116–120
XXVIII	$HO—CH_2—CH_2—S—CH_2—CH_2—S—CH_2—CH_2—O—CH_2—$ $CH_2—CH_2—CN$	118
XXIX	$NC—CH_2—CH_2—O—CH_2—CH_2—S—CH_2—CH_2—S—CH_2—$ $CH_2—O—CH_2—CH_2—CN$	110–112

The stabilizing agents in the third series are represented by

$$CH_2{=}CH—\overset{\oplus}{\underset{R}{S}}—CH_2—CH_2—OH \; X^{\ominus}$$

This class of compounds is synthesized by the alkylation of compound XXIV with a substance RX, where R may be an aliphatic, aromatic, or cycloaliphatic group and X^- is generally an anionic group to compensate the positive charge on the sulphonium bridge. Details of R and X^- are given in *Table 8*.

Compounds XXV to XXIX listed in *Table 7* and compounds XXX to XXXVIII listed in *Table 8* were also tested. The results of the induction time test are presented in *Figures 7* and *8*.

267

Table 8

No.	R	X^-	Solid point, °C
XXX	—CH_2—CH_2—OH	$HCOO^-$	
XXXI	—CH_2—CH_2—OH	$\frac{1}{2}COO^-$ | COO^-	35
XXXII	—CH_2—CH_2—OH	$\frac{1}{3}PO_3^{3-}$	
XXXIII	—CH_2—CH_2—OH	$\frac{1}{2}SO_3^{2-}$	40
XXXIV	—CH_2—CH_2—OH	$\frac{1}{2}H_2N$—CH$\genfrac{}{}{0pt}{}{SO_3^-}{SO_3^-}$	Decomp. at 70
XXXV	—CH_2—CH_2—OH	⟨C₆H₄⟩—O^- (X)	30–35
XXXVI	—CH_2—CH_2—OH	⟨C₆H₃⟩—O^- (X, X) X = *t*-butyl	40–45
XXXVII	—CH_2—CH_2—OH	CH_3—⟨C₆H₂⟩—O^- (X, X)	45
XXXVIII	—CH_2—CH_2—OH	Citric acid	65

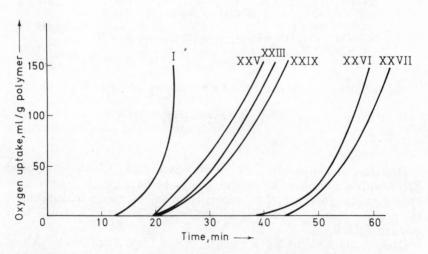

Figure 8. Induction times for general-purpose unstabilized polyethylene (I) and for polyethylene stabilized with inhibitors Nos. XXV to XXIX.

268

In *Figure 8*, the polyethylene (I) was stabilized with 0.1 % wt of compounds XXV to XXIX. The most efficient compounds are XXVI and XXVII, the mono-acid diesters of compound XXV, i.e.

$$HO—CH_2—CH_2—S—CH_2—CH_2—S—CH_2—CH_2—OH,$$

with phosphorous acid.

In *Figure 9*, the polyethylene (I) was again stabilized with 0.1 % wt of compounds XXX to XXXVIII. All the compounds are extremely efficient with the exception of XXX and XXXV, i.e. the formic and phenolic compounds.

The relationship of the proportion of functional group to time is presented in *Figure 10*.

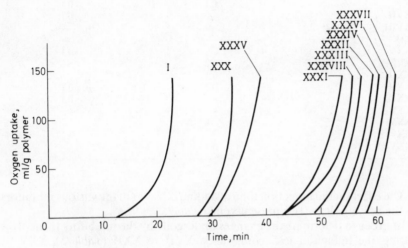

Figure 9. Induction times for general-purpose unstabilized polyethylene (I) and for polyethylene stabilized with inhibitors Nos. XXX to XXXVII.

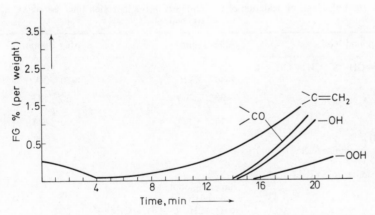

Figure 10. Proportion of the functional groups in polyethylene I stabilized with 0.1 % wt of XVII as a function of time at 180°C in the induction period test.

269

Table 9. Relationship between the type of stabilizer and functional group in oxidized polyethylene (I) (after ten minutes)

Type of stabilizer	$>CO$	$-OOH$	$>C=CH_2$	$-C=C-$	$-OH$ (% wt)
None	1.0	0.5	2.5	1.5	0.2
XVII	—	—	0.2	—	—
XVIII	—	—	—	—	—
XIX	—	—	—	—	0.1
XX	—	—	0.1	—	0.1
XXI	*ca.* 0.1	—	0.1	—	0.1
XXII	—	—	—	—	—
XXV	—	—	0.1	—	—
XXVI	—	—	0.1	—	0.1
XXVII	—	—	—	—	0.1
XXVIII	*ca.* 0.1	—	—	—	—
XXIX	—	—	—	—	—
XXX	—	—	—	—	0.1
XXXI	—	—	0.2	—	—
XXXII	—	—	0.1	—	—
XXXIII	—	—	0.2	0.1	0.1
XXXIV	—	—	—	—	0.1
XXXV	0.1	—	—	—	—
XXXVI	—	—	—	—	—
XXXVII	—	—	—	—	—
XXXVIII	—	—	0.3	—	—

The effect on the induction time of adding 0.1 % wt of the various inhibitors to the polyethylene (I) is presented in *Table 9*.

In order to determine the changes undergone by the inhibitors themselves during the induction test, compounds XVII to XXII (*Table 5*), XXV to XXIX (*Table 7*), and XXX to XXXVIII (*Table 8*), were oxidized at 180°C for ten minutes and twenty minutes. The results are listed in *Table 10*.

Table 10(a). Products of oxidation of the inhibitors in the induction time test (180°C in pure oxygen)

Compound No.	After 10 min	After 20 min
$CH_2=CH-S-CH_2-CH_2-R$ XVII	~40% XVII	~20% XVII
	~30% $>SO^*$	~35 $>SO$
	~10% $>SO_2^*$	~30 $>SO_2$
XVIII XIX XX } similar to XVII		
XXI similar to XVII	~5% dimer cyclobutane derivatives $CH_2-CH-S-CH_2-CH_2-R$ $CH_2-CH-S-CH_2-CH_2-R$	~2%

270

Table 10(a).—*Continued*

XXII	~35% XXII	~10% XII
	~10% $>$SO, P^{III}	~10% $>$SO, P^{III}
	~10% $>$SO$_2$, P^{III}	~25% $>$SO$_2$, P^{III}
	~20% P^V —S—	~30% P^V, —S—
	~10% P^V $>$SO	~20% P^V, $>$SO
	~10% P^V $>$SO$_2$	~5% P^V, $>$SO$_2$

* $>$SO groups or $>$SO$_2$ groups instead of S in the basic inhibitor.

Table 10(b). Products of oxidation of the inhibitors in the induction time test (180°C in pure oxygen)

Compound No.	After 10 min	After 20 min
XXV	20% XXV	15% XXV
	15% $>$SO	35% $>$SO
	10% $>$SO$_2$	20% $>$SO$_2$
	50% condensed polymers linear and cyclic (—CH$_2$—CH$_2$—S—)$_n$	30% condensed polymers
XXVI	30% XXVI	20% XXVI
	10% $>$SO, P^{III}	5% $>$SO, P^{III}
	5% $>$SO$_2$, P^{III}	10% $>$SO$_2$, P^{III}
	10% $>$SO, P^V	20% $>$SO, P^V
	20% $>$SO$_2$, P^V	35% $>$SO$_2$, P^V
XXVII	25% XXVI	20% XXVI
	20% $>$SO, P^{III}	20% $>$SO, P^{III}
	10% $>$SO$_2$, P^{III}	20% $>$SO$_2$, P^{III}
	20% —S—, P^V	25% —S—, P^V
	20% $>$SO, P^V	15% $>$SO, P^V
	5% $>$SO$_2$, P^V	10% $>$SO$_2$, P^V

XXVII ⎱
XXIX ⎰ not examined

Table 10(c). Products of oxidation of the inhibitors in the induction time test (180°C in pure oxygen)

$$CH_2{=}CH{-}\overset{+}{\underset{R}{S}}{-}CH_2{-}CH_2{-}OH \qquad X^-$$

The following were formed in all cases

(a) 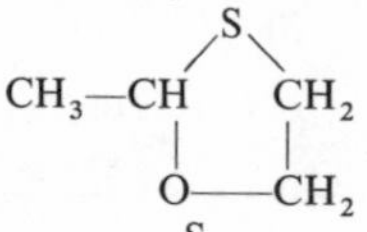 methylthioxolene $\sim0{\cdot}5 \to 2.5\%$

(b) XL 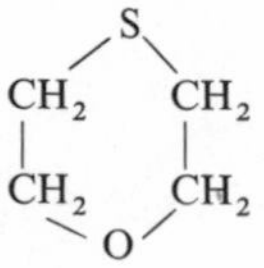 thiodioxane $\cdot 1 \to 3\%$

(c) XLI oligomers formed by intramolecular OH-addition to the vinyl double bond

$$|{-}O{-}CH_2{-}CH_2{-}S{-}CH_2{-}CH_2{-}|_{5-15} \sim 10\%$$

(d) Split compounds; from 2-mercaptoethanol (HS—CH$_2$—CH$_2$—OH), which was oxidized to HO$_3$S—CH$_2$—CH$_2$—OH, CO$_2$, SO$_2$ and lower carboxylic acids, and from HO—CH$_2$—CH$_2$—S—CH$_2$—CH$_2$—S—CH$_2$—CH$_2$—OH (XXV) about 3%, and from XXV itself, which was oxidized to $>$SO and $>$SO$_2$ compounds 2.5 to 5%.
XXV was formed by the addition of 2-mercaptoethanol (ME) to XXIV

$$HO{-}CH_2{-}CH_2{-}SH + CH_2{=}CH{-}S{-}CH_2{-}CH_2{-}OH \to XXV$$

ME   XXIV

X$^-$ decomposes as follows

No.	After 10 min	
XXX	CO, CO$_2$	30%, 30%
XXXI	CO$_2$	50%
XXXII	P^V	40%
XXXIII	SO$_2$	50%
XXXIV	NH$_3$, CO$_2$, SO$_2$	10%, 10%, 20%
XXXV	oligomeric phenols	10%
	alkylated phenols	10%
XXXVI } XXXVII }	similar to XXXV	
XXXVIII	CO$_2$	20%
	aconitic acid	20%

CALCULATION OF ELECTRON DISTRIBUTION BY THE HMO MODEL†

Electron distributions that are interesting for the reactivity are presented as $\Delta\rho$ values for αC, βC and acceptor R in *Figure 11*. The $\Delta\rho$ values are the differences between the individual nuclear charges and the π-electron densities.

†The electron distribution was calculated by Dr Feichtmayr, BASF.

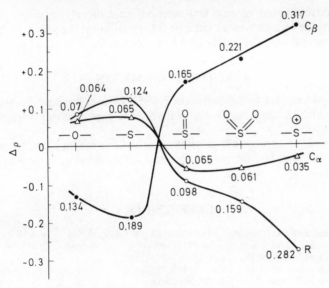

Figure 11. Electron distribution in the system
$$CH_2{=}CH{-}R-CH_2-CH_2-OH.$$

1 (βC)	1.13404	$\Delta\rho = -0.134$
2 (αC)	0.93643	$\Delta\rho = +0.064$
3 (Acceptor R)	1.92955	$\Delta\rho = +0.070$

The $\Delta\rho$ values for the other systems were calculated in the same way. The calculation confirms differences in behaviour of the various substituted sulphur derivatives in the experiment.

CONCLUSION

An approach combining analysis and synthesis of both the substrate—in this case polyethylene—and the additive has been adopted to obtain an insight into oxidative degradation by means of data that can be obtained by analysis. New inhibitor systems containing sulphur have been developed. These enter into competitive reactions and are themselves oxidized to inactive products, blocking chain reactions and decomposing peroxide groups.

Stabilization and the decomposition of polymers under defined conditions are problems that will remain topical as long as polymers are ingredients of articles that come into contact with food, etc. For this reason, optimum stabilization is always important.

In order to clarify the relationships between structure and active functional groups, a start has been made on the following studies:

(a) Checking the results obtained to see if they are valid for other systems too, e.g. polypropylene, polyisobutylene and polydienes;

(b) The development of new test methods that closely simulate practical conditions and give more realistic information on the value of the polymers in end use. This study also includes mechanochemical reactions.

ACKNOWLEDGEMENTS

My thanks are due to the staff of the BASF Polymer Research Laboratories for their valuable assistance in carrying out the experimental work required for this paper.

I also wish to express my gratitude to the Organizing Committee of the 11th IUPAC Microsymposium in Prague for their kind invitation to present this paper.

REFERENCES

[1] P. L. Hanst and J. G. Calvert, *J. Phys. Chem.* **63**, 104 (1959).
[2] B. K. Morse, *J. Amer. Chem. Soc.* **79**, 3375 (1957).
[3] M. S. Kharasch et al., *J. Org. Chem.* **15**, 763, 775 (1950).
   M. S. Kharasch et al., *J. Org. Chem.* **16**, 113 (1951).
[4] J. R. Thomas, *J. Amer. Chem. Soc.* **77**, 246 (1955).
   J. R. Thomas and O. L. Harle, *J. Phys. Chem.* **63**, 1027 (1959).
[5] M. H. Dean and G. Skirrow, *Trans. Faraday Soc.* **54**, 849 (1958).
[6] R. G. R. Bacon, *Quart. Rev. Chem. Soc. Lond.* **9**, 287 (1955).
[7] G. Scott, in *Atmospheric Oxidation and Antioxidants*, p 52. Elsevier: Amsterdam (1965).
[8] D. Barnard, *J. Chem. Soc.* 489 (1956).
[9] N. Kornblum and H. E. De La Mare, *J. Amer. Chem. Soc.* **73**, 880 (1951).
[10] C. E. Swift and F. G. Dollear, *J. Amer. Oil Chem. Soc.* **25**, 52 (1948).
[11] L. Baleman and K. R. Hargrave, *Proc. Roy. Soc. A*, **224**, 399 (1954).
   K. R. Hargrave, *Proc. Roy. Soc. A*, **235**, 55 (1956).
[12] G. Scott, in *Atmospheric Oxidation and Antioxidants*, p 275. Elsevier: Amsterdam (1965).
[13] J. L. Bolland, *J. Chem. Soc.* 1596 (1955).
[14] Yu. A. Shlyapnikov and V. B. Miller, in *The Aging and Stabilization of Polymers*, edited by A. S. Kuzminskii, p 36. Elsevier: London (1971).
[15] A. A. Berlin and S. J. Bass, in *The Aging and Stabilization of Polymers*, edited by A. S. Kuzminskii, p 162. Elsevier: London (1971).